Taunton's
# BUILD LIKE A PRO™
## Expert Advice from Start to Finish

# TRIM Carpentry
# and BUILT-INS

Expert Advice from Start to Finish

# TRIM Carpentry
# and BUILT-INS

## CLAYTON DEKORNE

The Taunton Press

# The Taunton Press
Inspiration for hands-on living®

The Taunton Press, Inc., 63 South Main Street, P.O. Box 5506, Newtown, CT 06470-5506

e-mail: tp@taunton.com

Distributed by Publishers Group West

COVER AND INTERIOR DESIGN: Lori Wendin

LAYOUT: Jeff Potter/Potter Publishing Studio

ILLUSTRATOR: Mario Ferro

Taunton's Build Like a Pro™ is a trademark of The Taunton Press, Inc., registered in the U.S. Patent and Trademark Office.

Printed in the United States of America

10 9 8 7 6 5 4

The following manufacturers/names appearing in *Trim Carpentry and Built-Ins* are trademarks and servicemarks: Accuride® 3017, Amana Tool® Corporation, Andersen® windows, Blum®, Bondo®, Bosch® DWM40L Miter Finder, Cabinet Factory®, CMT® USA, Inc., Collins Quality Tool®, Color Putty Co.®, DeWalt®, Disston®, Eagle® windows, Formica®, Fuller® countersink bits, Georgia-Pacific® Fiber Ply, Hartville Tool®, Jesada Tools®, Liquid Nails®, Minwax®, National Hardwood Lumber Association℠, Phenoseal®, Quik-clamp®, Rockler®, Roto Zip®, Scotch® tape, Speed Rip Square™, Stanley®, Surform® plane, Titebond®, Titus® connector, Wilsonart®, Woodcraft®, Woodworker's Supply, Inc.℠

***About Your Safety:*** Construction is inherently dangerous. Using hand or power tools improperly or ignoring safety practices can lead to permanent injury or even death. Don't try to perform operations you learn about here (or elsewhere) unless you're certain they are safe for you. If something about an operation doesn't feel right, don't do it. Look for another way. We want you to enjoy the craft, so please keep safety foremost in your mind whenever you're in the shop.

Library of Congress Cataloging-in-Publication Data

DeKorne, Clayton.

   Trim carpentry and built-ins : expert advice from start to finish / Clayton DeKorne.

        p. cm. -- (Build like a pro)

Includes index.

   ISBN 1-56158-478-9

   1. Trim carpentry--Amateurs' manuals.  2.  Built-in furniture--Amateurs' manuals.  I. Title.  II. Series.

   TH5695 .D45 2002

694'.6--dc21                                        2002004331

For my daughters, Helen and Cecelia, who waited patiently while I was hunched over the computer, saying far too many times, "Just a minute, I'll be right there."

# Acknowledgments

TO WRITE THIS BOOK, I am indebted to both of my fathers: Jim DeKorne, who taught me to be a good mechanic in every trade I've undertaken, including writing, and Rick McKinney, whose unconditional love for me has been a constant source of renewal. I am also deeply grateful to J. Ladd, who shared his carpentry expertise and shaped my practice in the trade.

I could never have completed this book without the loving support of Robin Michals, whose partnership over the last few years has sustained me more than words can say. I also greatly appreciate the excellent friendship of David Crosby, Christopher Pierson, John Wagner, and Pete Young, as well as the inspiration to do great work, as shown by Karel Bauer, David Dobbs, Jen Mathews, and Steph Pappas.

To complete this book, special thanks goes to Carolyn Bates and Andrew Kline, whose photographs provide far more insight into the work presented than my words ever could, and to Andrew Wormer, who provided editorial guidance and supreme patience for the duration of this project.

Finally, I wish to thank the many carpenters, contractors, building suppliers, and other trade experts who shared their knowledge and never hesitated to give me their time and attention. Chief among them are Larz Allen, Tom Morse, Clark Sargent, Craig Tougas, Paul and the rest of the crew at Gregory Supply, and Dave and Bruce of Sterling Hardwoods. In addition, I owe great appreciation to a number of excellent teachers of the trades, including Sal Alfano, Charles Berliner, Butch Clark, Jed Dixon, Don Dunkley, Steve Farrell, Carl Hagstrom, Will Holladay, Mark Luzio, Craig Savage, and Dave Severance. If you ever get the chance to be on a crew with any of the characters I've mentioned on this page, count yourself extremely lucky.

# Contents

■ CHAPTER ONE

## Project Planning   6

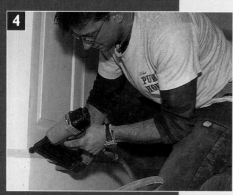

■ CHAPTER TWO

## Trim Materials   18

■ CHAPTER THREE

## Trimming Windows   30

■ CHAPTER FOUR

## Running Baseboard   50

# Introduction

THIS BOOK IS WRITTEN for anyone interested in becoming a good carpenter. I have set out to provide readers with a sense of what to expect when they gather the tools and materials to undertake an interior trim job. I have poured into these pages my insights about how materials behave over time, collected design details that will look good over the long term, and disclosed many trade tips that other professional carpenters have shared with me during the past 25 years.

Carpentry books often fail, I think, when they try to cover every conceivable way to accomplish a task. The result is a mass of information that is boiled down into neat but irrelevant categories or tucked into bland generalities that wander away from the actual experience of completing a carpentry job. In this volume, I have tried to avoid giving too much general information. Instead, I've focused on a few methods that have worked well for me as a professional trim carpenter.

Trim carpentry, as I address it here, refers to any kind of interior woodwork in a house, including door and window casings, baseboards, crown and ceiling moldings, wainscoting and other wooden wall paneling, cabinets, and built-in furniture. This book covers most types of "finish" woodwork—the carpentry details that are addressed before the painting and decorating begins. This book does not include specific details about the installation of doors and windows (which are better handled as part of the building envelope) or stairways (which are complex enough to comprise an entire book).

Overall, this edition focuses on the mechanics of how trim fits together. However, I feel strongly that no carpentry work can ever be separated from design or from building science. Carpenters must constantly make aesthetic decisions concerning proportion, scale, texture, color, and pattern, and like every other aspect of the trade, successful trim design is the result of conscious, informed effort, not accident. Equally important, carpentry is always closely tied to the physical properties of wood, which govern how it behaves in a changing environment. Wood trim mechanics must always be addressed from the perspective of dimensional stability. How much a board moves dictates how tight a joint will remain.

With this in mind, I hope that anyone interested in becoming the best carpenter possible strives to become a student of the house building trades, including mechanics, aesthetics, and science in equal measure. When you understand carpentry, there are no strict boundaries among these perspectives—they are each part of a whole way of thinking that is inseparable from the actual work. Above all, carpentry requires a way of thinking about the constructed world that cannot be learned in a book. At some point, a reader must pick up the tools and actively work with the materials. At that point, I hope the principles and methods described here will make that practice a richer experience.

As you read this book, bear in mind that I have written it from the perspective of a professional carpenter, adopting a "trade" perspective that equally values production and quality. Quality is always a relative term. You can go nuts with quality in carpentry. I have done jobs for customers who actually inspected miter joints with a magnifying glass and for others who didn't particularly care what the joinery looked like, as long as they could list "natural hardwood trim" in a rental advertisement. Doing each job "well" meant discerning completely different levels of quality.

The balance between production (getting the job done as efficiently and inexpensively as possible) and quality (executing it as elegantly and precisely as possible) sets a baseline for building practice. It's a baseline that works equally well for aspiring carpenters who wish to pursue the trade for its own sake and for homeowners who want the work they do to last for the next generation. That said, I feel confident that the methods described in this book will allow you to satisfy that person holding the magnifying glass, be it yourself, your supervisor, or your customer. But more important, the methods described here will allow you to get the work done.

# How to Use This Book

IF YOU'RE READING THIS, you're a doer who is not afraid to take on a challenging project. We designed this book and this series to help you do that project smoothly and cost effectively.

Many doers jump in and do, reading the directions only when something goes wrong. It's much smarter (and cheaper) to start with the knowledge of what to do and plan the process step by step. This book is here to help you. Read it. Familiarize yourself with the process you're about to undertake. You'll be glad that you did.

## Planning Is the Key to Success

This book contains information on designing a project, choosing the best options for the result you want to achieve, and planning the timing and execution. We know you're anxious to get started on your project. Take the time now to read and think about what you're about to do. It will help you refine your ideas and choose the best materials.

You'll find advice here on where to look for inspiration and how to make plans. Don't be afraid to make an attempt at drawing your own plans. There's no better way to get exactly what you want than by designing it yourself. But be honest with yourself. Seek the advice of an architect or engineer if you need it.

After you've chosen the project you want to undertake, make lists of materials and a budget for yourself, both in money and in time. There's nothing more annoying than a project that goes on forever.

## Finding the Information You Need

We've designed this book to make it easy to find what you need to know. The main part of the book details the essential parts of each process. If it's fairly straightforward, it's simply described. If there are key steps, these are addressed one by one, usually accompanied by drawings or photos to help you see what you will be doing. We've also added some other elements to help you understand the process better, find quicker or smarter ways to accomplish the task, or do it differently to suit your project.

### Alternatives and a closer look

The sidebars and features included with the main text explain aspects in more depth and clarify why you're doing something. In some cases, they describe a completely different way to handle the same situation. We explain when you may want to use that method as well as its advantages. Photos or drawings to help you visualize the text usually accompany

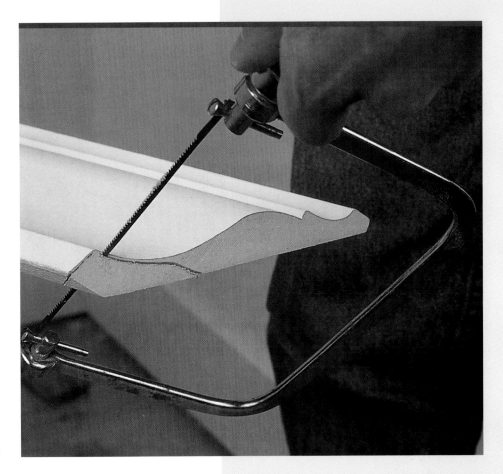

the sidebars. The sidebars are meant to help, but they're not essential to understanding or doing the process.

### Heads up!

We urge you to read the "Safety First" and "According to Code" sidebars. "Safety First" gives you a warning about hazards that can harm you. Always work safely. Use appropriate safety aids and know what you're doing before you start working. Don't take unnecessary chances. If a procedure makes you uncomfortable, try to find another way to do it. "According to Code" can help you avoid problems with your building inspector, building an unsafe structure, or having to rip your project apart and build it again to suit local codes.

### There's a pro at your elbow

The author of this book, and every author in this series, has had years of experience doing the kinds of project described here. We've put the benefits of their knowledge into quick "sound bites," which always appear in the left margin. "Pro Tips" are ideas or insights that will save you time or money. "In Detail" is a short explanation of a particular aspect of the job that may be of interest to you. While not essential to getting the job done, these are meant to explain the "why."

Every project has its surprises. Since the author has encountered many of them already,

he can give you a little preview of what they may be and how to address them. And experience has also taught the author some tricks that you can only learn from a pro. These include tips, tools and accessories you can make yourself, or materials and tools you may not have thought to use.

# Building Like a Pro

To make a living, a pro needs to work intelligently, quickly, and economically. That's the strategy presented in this book. We've provided options to help you make the best choices in design, materials, and methods. That way, you can adjust your project to suit your skill level and budget. Good choices and good planning are the keys to success. And remember: All of the knowledge and skills that you acquire while working on a project will make the next one easier.

# Project Planning

# CHAPTER ONE

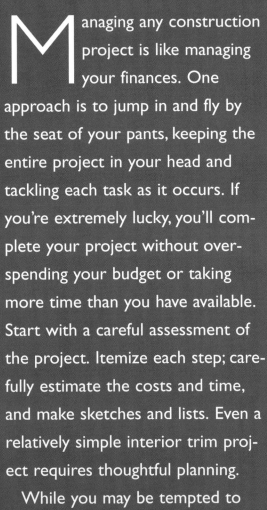

Managing any construction project is like managing your finances. One approach is to jump in and fly by the seat of your pants, keeping the entire project in your head and tackling each task as it occurs. If you're extremely lucky, you'll complete your project without overspending your budget or taking more time than you have available. Start with a careful assessment of the project. Itemize each step; carefully estimate the costs and time, and make sketches and lists. Even a relatively simple interior trim project requires thoughtful planning.

While you may be tempted to skip this chapter to get to work, take the time to understand the importance of creating the key development "documents" that go into planning a project. Do this before you start, and the project will begin to take shape in your mind and on paper. The reward will be efficient, smooth construction and a great finished result.

### TRADE SECRET

One essential "tool" I always keep close at hand is a stack of blank paper (usually recycled from the backs of used printer paper) on a clipboard, with a big postal rubber band around the base to keep the pages from flapping. This sits on my workshop bench or near my toolbox when I'm on site. I use it to sketch the myriad little details that I have to think through in the course of any finish job.

### TRADE SECRET

If you're building cabinets or built-ins, lay out a full-scale plan of your proposed project on the floor. Use a lumber crayon to mark the cabinet positions directly on the subfloor. If you're building on a finished floor, tape down rosin paper (sold in rolls at lumberyards for laying under hardwood floors) and mark your drawing on that.

# Sketching a Design

**Interior trim projects** can range from the relatively simple to the relatively complex. Even if you just run baseboard or case a window, you'll want to consider such questions as: How will this new work relate to the existing house? How will the horizontal planes of my window trim line up with the horizontal lines of the room? How will the vertical lines relate to each other? Does the ornamentation of these moldings fit the design period of the house? How far will this rail stand out from the wall? Will it stand out farther than the window casing, and if so, how will I finish the ends? These are the kind of questions you need to resolve at the design phase of a project.

For me, the easiest way to work out a design is to make a series of drawings. I think best when I can see what I'm thinking about. So from the outset of any project, I take up a pencil and start sketching.

For some projects, such as running baseboard, it's not necessary to go through all the drafts I define here—a simple conceptual floor plan, with dimensions for each wall, will usually be enough to give you the information you need to make a cut list and get started. However, if you're unsure about the height of the baseboard in proportion

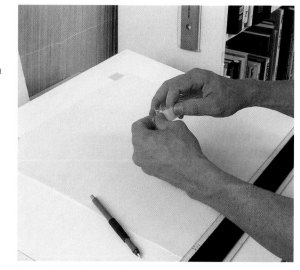

A simple drafting setup—a drawing board, a sliding square, and vellum held down with drafting tape—makes it much easier to sketch ideas and create working drawings.

to other elements in the room, you may want to draft a detailed elevation that situates the baseboard in relation to the height of the doors, windows, and ceiling.

More complex projects call for more detail. A window stool should be carefully drawn in plan so you know exactly how wide and how long to cut the horns. You may want to draw a scaled elevation of wainscoting so that you can envision how high the chair rail will be and think through how it intersects with windows and doors. And a full

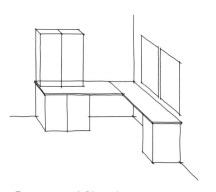

**Conceptual Sketch**
*When planning a project, begin by sketching the basic items that you intend to build. The drawing need not be elegant. At this point, you are concerned with only the general massing of components and the possible correspondence of shape to existing features, such as a window.*

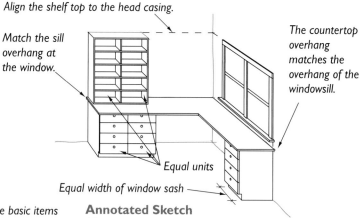

*Align the shelf top to the head casing.*

*Match the sill overhang at the window.*

*The countertop overhang matches the overhang of the windowsill.*

*Equal units*

*Equal width of window sash*

**Annotated Sketch**
*After an initial sketch has defined the components, add detail by thickening the edges and working out critical alignments. Include notes that will help define the finished measurements.*

wall of bookshelves or a built-in entertainment center calls for a complete series of sketches to work out everything from the massing of upper and lower units to the shelf intervals for books, electronic equipment, and even art objects. For more involved projects, I begin with a set of conceptual models, and then draft a set of detailed drawings that show the project in different planes. These preliminary drawings finally evolve into dimensioned drawings from which I actually pull off construction measurements. This three-phase progression can result in one final sketch or a ream of drawings that together comprise a carefully considered approach to the project.

## Step 1: Conceptual sketches

The most important part of the design process is the early conceptual stage. Don't worry too much about drawing to scale at this point. If you don't feel very practiced, you may want to draw on vellum over graph paper. However, the grid on graph paper should only be used as a rough measure for casting things proportionally. Don't get hung up on the precision of the grid. Focus more on getting a look and feel. Conceptual drawings should focus on form, not detail. They are a way of defining the basic shapes and carving out the space that the new project will occupy.

I start with a simple perspective view. Even if you're not practiced in all the rules of perspective drawing, you can sketch out simple 3-D boxes that can be developed to show the massing and shapes of your project. Play with the shapes, and if you're feeling bold, add shadow. Don't overdo this—it should be fast. But it will help you think about how the solid form fits into the space of your home. Does a cabinet project too far out from the wall? What is the proportion between the height of the baseboard and the ceiling height of the room? How wide should the windowsill be to pick up the built-up casing? The more realisti-

# The Basic Box in Perspective

Sketching in 3-D perspective doesn't have to be an elaborate process. Many of us learned how to do this as children by drawing overlapping boxes and joining them at the corners with lines, as shown here in the Method I series.

Once you get the feel for drawing a basic box, you can sketch the front face of a box and project lines off the corners at an angle. Finish the sketch by drawing the "back" lines parallel to the top and sides.

After you're comfortable drawing a basic box without overlapping lines, you can create more complex sketches by stacking a series of boxes to represent the base and upper cabinet units, as shown.

**Method 1: Getting the Idea**
*Drawing volumetric cubes evolves from the trick many of us learned in grade school.*

*1. Start with a square.*

*2. Draw a second square overlapping at one corner.*

*3. Connect the corresponding corners.*

**Method 2: Learning to Sketch**
*The process of drawing volumetric shapes can be speeded up by sketching.*

*1. Start with a square.*

*2. Give it dimension.*

*3. Limit the dimensions.*

*4. Clean up the lines.*

**Stacking Boxes**
*When sketching interior cabinets, think in "boxes" of various sizes. These will actually be similar to the units that you will build.*

 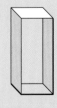 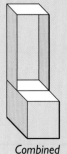

*Short, squat (base unit)*

*Narrow, tall (shelf unit)*

*Combined*

## PRO TIP

*A drawing board can be as simple as plywood propped up on a 4×4 block. Make sure you keep the surface of the board smooth and clean.*

### IN DETAIL

I use a plain HB drawing pencil for most lines and a softer B pencil for bold outlines. For erasers I prefer artist's kneading rubber. This material doesn't abrade paper or vellum. Graphite sticks to it, so it lifts off pencil marks and daubs away smudges.

### TRADE SECRET

High-quality vellum won't disintegrate as you erase. Most vellum is translucent, too, so you can see a piece of graph paper beneath it or trace over a previous draft of your drawing. Use drafting tape, not masking or Scotch® tape, to secure the drawing to the drawing board.

**Besides its traditional application, wainscoting can also be used to provide interesting details in bathrooms.**

cally you can see these details beforehand, the smoother the project will go.

Those who are inclined to doodle on scrap paper will be more comfortable with making conceptual drawings. For others, it will prove frustrating to draw in a loose style that doesn't look anything like what you have in mind. If this is the case, you should still make a stab at Step 2, creating preliminary plan and elevation views of the project.

## Step 2: Preliminary drawings

The purpose of preliminary drawings is to create a series of pictures that accurately portray the finished project. Each of these pictures represents a different plane—the bird's-eye view that gives you a *plan,* the imaginary slice through the work that yields a *section,* and the flattened front view that results in an *elevation.*

At this stage of drawing, don't worry too much about dimensions. Instead, include furniture and lines that represent the existing structure (save dimensions for the final working drawings). By including furniture, fixtures, and wall lines, you can begin to judge the size of the project in the context of the room where it will be built. Also, experiment with line weights. For example, draw the project's outlines in a heavy dark line, and then fill in all double-line walls. These heavy lines make it easier to see the layout at a glance.

## Step 3: Measured working drawings

Construction drawings (or working drawings) should be drawn to scale and include dimensions written on the paper. I use an architect's scale to help me make these drawings. Most of the time, I choose a ¼-in.:1-ft. scale, unless I need to work out close-up details.

At this point, you need to know exact measurements. If you are installing window casing, you must know the exact window dimensions. If you are building an entertainment center, you need to know the exact dimensions of the television set, estimate the number of videos you want to store,

**A T-square, which slides on the edge of a drawing board, and a triangle, which rides on the edge of the T-square, allow you to draw right angles instantly anywhere on the page.**

and know the size of any stereo equipment that will be included. You should also account for the exact locations of electrical switches and outlets, heating ducts, and any other house features that intersect with your project.

For any large project, dimensioned construction drawings are a must. If your trim work is part of a larger project that requires a building permit, an inspector will insist on seeing dimensioned drawings. Similarly, if you need to hire an electrician or a plumber to move wires or pipes, it's good to have clear construction drawings when discussing the estimate. It will inform the subcontractor of exactly what you want and may even save you money if the sub doesn't have to guess at unknowns and pad the price to cover them.

## Only a map

Any sketch or detailed drawing only provides an ideal map. When you walk into the room that you will trim out, the floor and walls won't be composed of the perfect rectangles and right angles you have just drawn. And the material you use will have dimension, differing markedly from the plans and elevations you have drafted.

Despite these limitations, however, the drawings serve an all-useful organizing purpose. I keep the drawings with me on a clipboard, along with plenty of blank paper for making notes and revisions. I start making notes to myself in the margins of the drawing and later expand those notes into material and cut lists.

Notes can also serve as useful guides for executing the work. For example, I might draw an arrow to a corner of the room on my drawing and note which way the corner is out of square or how much the wall is out of plumb. That way, I start to compile a record of actual conditions, which I can carry with me to the saw, helping me remember which way to compensate when I cut the baseboard to fit in that out-of-square corner.

## Scaling Drawings

Scale refers to both the ruler and the proportions used by carpenters and architects to create measured drawings. As proportions, a scale expresses a precise relationship between a design drawn on paper and a full-size object.

Scale is expressed numerically as a proportion—for example, ¼:1— which means that every ¼ in. measured on a drawing equals 1 ft. measured in physical space. On a scale ruler, every fractional inch— ¼, for example—is labeled 1, 2, 3, and so on, indicating a number of feet. Common architectural scales used for working drawings are ¼:1, ½:1, 1:1, and 3:1. In each case, the first number always implies a measurement in *inches,* the second always means *1 foot.*

The scale you choose for measured drawings depends on the amount of detail you need to show. For most working drawings of large areas—an entire wall of cabinets or the floor plan for a baseboard layout, for example—I typically draw in ¼:1 scale (¼ in. = 1 ft.). For more detailed work, I use a 1:1 scale (1 in. = 1 ft.). For complex details, however, it often helps to magnify the drawing with a 3:1 scale (3 in. = 1 ft.).

Scale rulers come in a variety of shapes and sizes. Make sure you use an *architect's scale,* which is calibrated in ³⁄₃₂ in.-, ³⁄₁₆ in.-, ⅛ in.-, ¼ in.-, ⅜ in.-, ¾ in.-, ½ in.-, 1 in.-, 1½ in.-, and 3 in.-per-ft. increments. Don't use an *engineer's square,* which is divided into 10, 20, 30, 40, 50, and 60 parts of an inch.

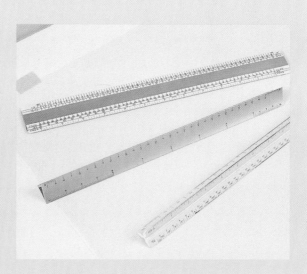

**Three scales (top to bottom): All of these are "architect's" scales, which show proportional distances divided in feet and inches (see the close-up below). When buying one, make sure you don't pick up an engineer's scale, which may look identical but divides the scales into decimal equivalents (divisions of 10).**

**On an architect's scale, the first ¾-in. increment is broken into 12 divisions for laying out scaled inches.**

## TRADE SECRET

One of the easiest ways to clarify an architectural drawing is to vary the line weights. Use the darkest, boldest line for outlines, lighter lines for details (such as drawer and shelf demarcation or millwork profiles), and dashed lines for underlying parts of an assembly. In some cases, it may help to heavily mark repeating units as a way of defining the piece's mass.

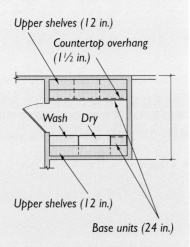

Upper shelves (12 in.)

Countertop overhang (1½ in.)

Wash   Dry

Upper shelves (12 in.)

Base units (24 in.)

**Laundry room storage**

**Vary Line Weights**
*To help clarify a drawing, vary the line weights to distinguish the different elements.*

**This clever use of built-ins in a child's room has pull-out drawers that create seating.**

# Material Take-Offs

Once you have a workable set of dimensioned drawings, you are ready to do a material take-off. This estimate not only serves as a shopping list but also helps you categorize the distinct tasks involved in executing the work. As mentioned above, I begin making notes in the margins as I develop my drawings. These notes relate not only to the inconsistencies in the actual dimensions of a room (which corners are out of square, where the floor or ceiling dips, where the drywall stands out from the window jamb, and so on) but also to the specifications for the project. This is a good place to begin keeping track of all the little decisions that go into a construction project: What is the wood choice? How wide is the casing? How thick is the window stool? What hardware (for example, hinges, drawer slides, catches, and pulls) will be needed?

When working up a design, you have already made a lot of decisions about the materials you need. By transferring the margin notes to a list, you have the beginnings of a material list. But at this point, the approach to generating such a list has been somewhat haphazard, consisting of those items that have jumped out at you as you work through the design. What's needed now is a logical take-off method that allows you to avoid wasting money by buying too much stock, or wasting time by not buying enough, and having to go back and forth to the lumberyard and hardware stores at inopportune moments.

## Trim take-offs

As the name implies, a take-off is taken off the drawings. For trimwork, I organize my list in two categories—standing trim and running trim.

*Running trim* refers to all the continuous horizontal moldings, such as baseboard, shoe molding, base cap, crown molding, and chair rail. This material is typically sold by the linear ft. and often ordered in random lengths. When generating a take-off for the amount needed, I work off the floor plan, totaling the lengths of the walls and writing down a linear-ft. measurement. I typically don't subtract for doors when estimating baseboard. Similarly, I don't subtract for windows and doors, unless there is a large bank of windows or an extra-wide door opening. And I add 10% to 15% extra onto solid-wood stock to allow for waste from defects, miscuts, and miters. For MDF materials, which generally have very few defects, I only need to add 5% to 10% extra.

*Standing trim* refers to all of the window and door casings and sometimes the wainscoting materials. It's typically ordered in specific lengths and sold in lengths in multiples of 2 ft. In an entire house estimate, standing trim is often sold in packages for each door and window, and I find this makes it easy to calculate, even if I am just trimming out a few windows and doors. A package consists of the trim needed to case one side of a door or window. For each package, I figure out how much I need, and then translate it into standard-length boards.

For example, one side of a standard 6-ft., 8-in.-tall by 2-ft., 6-in.-wide door requires two 7-ft. lengths for the side casings and one 3-ft. length for the head casing (the extra inches allow room for the miters and splits or checks in the ends). To make ordering easy, I call for one 10-ft. piece and one 8-ft. piece. However, if I have several doors to do, I often specify 14-ft. material for all of the side casings (one length for one door's two side casings) and gang up all the head casings into one board, such as a 10-ft. piece for three doors (with 1 ft. of waste) or a 12-ft. piece for four doors.

Windows can be tallied in a similar manner, except you need to double up on the header (one length is actually the apron) and include a separate item for the stool (usually made with 5/4 stock instead of regular 1× stock). Overlay moldings on built-up casing is an exception. Since it is often composed of molding stock that's sold in random lengths by the linear foot, I tally the perimeter dimensions of the windows and write down the number of linear ft. needed, as with running trim.

## Sample Take-Off

| Rm | Description | Material | Lengths | Total LF | Unit cost | Cost |
|-----|-------------|----------|---------|----------|-----------|------|
| BRM | BASEBOARD | RANCH BASE/PINE | 2-12' 3-10' | 45 | | |
| HALL | " | " | 1-16' 1-12' | 30 | | |
| | | | | | | |
| BRM | DOOR CASE | 1x4/PINE SELECT | 3-14' 1-12' | | | |
| BRM | WINDOW CASE | " | 2-8' 2-10' | | | |
| BRM | WINDOW STOOL | 5/4 x6/PINE SEL. | 1-10' | | | |
| BRM | DOOR MOLD. | 1/2 x 1 HALF RD. | RANDOM | 60 | | |
| BRM | WINDOW MOLD. | " | " | 40 | | |

Project: TRIM FOR NEW ADDITION   Date:

Kitchen cabinets are the most common form of built-ins. In this breakfast nook, built-in bench seats conceal extra storage space.

## Cabinet take-offs

For cabinets and built-ins, I use three material categories: solid stock, sheet stock, and hardware.

*Solid stock* includes all of the face banding, edge banding, fascia, scribe strips, and hanging rails. I calculate these much like I do for standing trim, as the items are typically cut from standard-dimension

## PRO TIP

*Try to think of all the small things you may need. Stopping in the middle of work to buy something you forgot is frustrating and wastes time.*

### IN DETAIL

Adhesives can be useful problem solvers. For example, Liquid Nails® comes in handy for holding cabinet rails where studs are missing, and a similar type of subfloor adhesive can batten down loose floor boards beneath a cabinet. My all-time favorite for securing paint-grade trim is vinyl adhesive caulk. I use it along with finish nails to affix cabinet facing, base caps, overlay moldings on casing, and chair rails.

### WHAT CAN GO WRONG

Don't use wood glue to fill gaps caused by sloppy craftsmanship and expect a lasting bond or a clean-looking joint. It's better in the long run to cut the joint airtight, both for strength and for appearance.

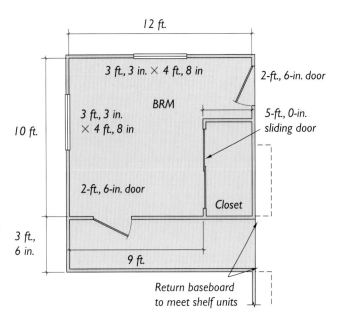

**Baseboard Plan**

*When working out a trim take-off, pull dimensions off the plan, converting each length to standard lumber lengths.*

| BRM | | HALL | |
|---|---|---|---|
| 2—10 ft. | | 1—16 ft. | |
| 2—12 ft. | | 1—16 ft. | |
| −LF | = 44 ft. | −LF | = 28 ft. |
| +add | = 4 ft. | +add | = 2 ft. |
| Total LF = 48 ft. | | Total LF = 30 ft. | |

| DOOR CASING | | WINDOW CASING | |
|---|---|---|---|
| 3—14 ft. | | 2—8 ft. | |
| 1—12 ft. | | 2—10 ft. | |
| Total LF = 54 ft. = 60 | | Total LF = 36 ft. = 40 | |

STOOL    1—10 ft.

lumber. However, many of the rails and stiles for cabinets are ripped from wider stock. For example, $1\frac{5}{8}$-in.-wide face banding and hanging rails can be cut from 1×4s (yielding two lengths per board). However, $2\frac{1}{4}$-in.-wide rail stock should come from a 1×8, which provides three lengths per board and leaves only about $\frac{5}{8}$ in. of waste (always account for a $\frac{1}{8}$-in. kerf when figuring your rips).

*Sheet stock* includes all of the cases, shelving, door panels, and (often) drawer fronts. I calculate these by the sq. ft., rounding up to the nearest ft. To simplify this, first calculate the sq. ft. of a cabinet's side, bottom, and doors, and then repeat these numbers for each instance. The square footage number you obtain must be converted to a number of plywood sheets. To do this, divide

## Estimating Plywood for Sample Project

**Base Unit = 24 in.×30 in.×36 in.**

| ¾-in. plywood | a) Sides | 2 @ 30 in.×24 in. = 10 sq. ft. |
|---|---|---|
| | b) Bottom | 1 @ 36 in.×24 in. = 6 sq. ft. |
| | c) Doors | 2 @ 18 in.×30 in. = 4 sq. ft. |
| ¼-in. plywood | d) Back | 1 @ 30 in.×36 in. = 8 sq. ft. |

**Shelf Unit = 12 in.×40 in.×36 in.**

| ¾-in. plywood | e) Sides | 2 @ 12 in.×40 in. = 7 sq. ft. |
|---|---|---|
| | f) Top and bottom | 2 @ 12 in.×36 in. = 6 sq. ft. |
| | g) Shelves | 4 @ 12 in.×36 in. = 12 sq. ft. |
| | | (yields 2 extra shelves) |
| ¼-in. plywood | h) Back | 1 @ 40 in.×36 in. = 10 sq. ft. |
| | | Total ¾-in. plywood sq. ft. = $\frac{45}{32}$ = 2 sheets |
| | | Total ¼-in. plywood sq. ft. = 18 = 1 sheet |

*Note:* This is the amount of plywood needed for *one* base and *one* upper cabinet as shown in the drawing on the facing page.

by 32, the number of square feet in a standard 4×8 sheet of plywood. Here again, I round up to the next highest number to allow for some waste.

The most difficult part of a cabinet take-off is dividing the plywood so that cabinet parts are cut from it efficiently. To clarify this process, make a drawing that shows each part you need. From that, generate a materials list, indicating a rough estimate of the sq. ft. needed for each component and the number of sheets needed for the project. A quick sketch of a sheet of plywood will help you plan how to cut each component from that sheet. For example, this built-in (see the drawing at right) requires four sheets of ¾-in. plywood and two sheets of ¼-in. plywood for the two lower units and the two upper units that comprise it.

*Hardware* includes all of the hinges, drawer slides, door pulls, and shelf pins, as well as any catches, bumpers, and specialty hardware, such as TV extensions or tape holders. To figure this out, I mark the elevation drawings with an X at each hinge location, a circle at each pull location, and a circled X at each slide location. To calculate shelf pins, simply count the number of shelves you plan to install and multiply by four.

With cabinets, I also include 2×4s for kick bases. I tally these by counting 2¼ ft. of 2×4 for every 1 ft. of cabinet run. This allows for a cross-piece in the ladder base approximately every 4 ft.

## Supply lists

At the bottom of every list, I include a final category: supplies. This includes all of the fasteners (nails and screws), glues and adhesives, biscuits and dowel stock, caulk, veneer tape, finishes (paints and stains), sandpaper, steel wool, wood filler, shims, extra drywall compound for repairs, and whatever else may be needed. Many of the miscellaneous items are the hardest to forecast, and they are often the things for which I go running out to a hardware store in the middle of a project. Any time

**Cabinet Project Plan**

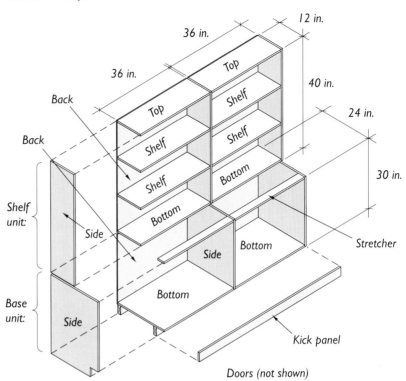

**Plywood Division**

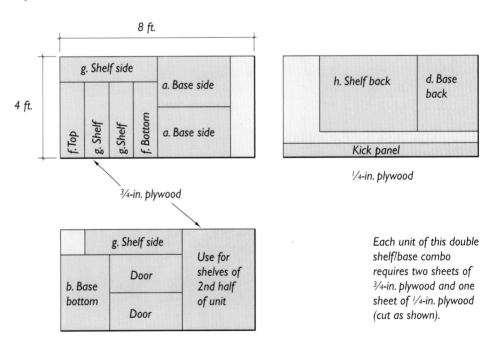

*Each unit of this double shelf/base combo requires two sheets of ¾-in. plywood and one sheet of ¼-in. plywood (cut as shown).*

## TRADE SECRET

If you're working on a traditionally styled project, you may find valuable information on trim styles and built-ins from books on historical homes. Also, nothing beats looking at actual trim profiles as a way of visualizing how the finished project will look and thinking of creative ways to combine components for a totally new look. An hour spent browsing the moldings at a lumberyard or mill shop may be just what you need to get your creativity in gear.

## TRADE SECRET

It never hurts to ask. In my experience, most counter agents at lumberyards and home centers willingly extend a wealth of information, making suggestions on alternative materials, correcting oversights on your materials lists, and disclosing details of how they package and price particular items.

one of these miscellaneous items occurs to you, write it in the margin notes of your drawing. You are most likely to think of them as you build the project in your mind and as you develop the drawings. Also, over time, you will acquire a well-stocked shop with many of these items.

# Pricing Materials

Once you have a complete materials take-off, you're ready to shop at a lumberyard or home center. At most places, you can present your lists at the counter and have it priced out (or, in many cases, you can fax your list if you already know the selling agent). Explain that you are getting an estimate of cost figures to compare to your budget. Some places may give you a price list and let you fill in the blanks, but this is increasingly rare and not often done, unless you are a contractor buyer (in which case, you may be given password access to an online price list and purchase manager).

## Price vs. value

It's always worth comparing prices at different suppliers, but I will caution you from buying solely on price. Sometimes (though certainly not always) the least expensive may not be the best value. In my area, for example, I have a great hardwood dealer who offers an incredible selection, custom milling services at a reasonable price, and valuable knowledge about the woods I seek. By comparison, the largest home center

## Sample Supply List

| Item | Size/type | Quantity |
|---|---|---|
| Shims | Cedar or pine | I bundle |
| Screws | 2 in. or 2½ in. (for hanging rail) | I lb. |
|  | 1⅝ in. (for case assembly) | I lb. |
| Nails | 6d finish (for face frame, shelf banding, and end panels) | I lb. |
|  | 4d ring-shank panel (for case backs) | I lb. |
|  | ¾-in. finish brads (for door edge banding) | I sm. box |
| Hinges | Blum® self-closing #91A6630 | 4 |
| Knobs | Blum mounting plate #195H7100 | 4 |
| Door bumper pads | 1¼-in.-diameter maple | 2 |
|  | ⅜-in.-diameter clear | 4 |
| Shelf pins | Spoon, Miller #306 | 16 (8 extra) |
| Glue | Type II | I pt. |
| Biscuits | #20 (for case assembly) | 26 (min.) |
| Caulk | Phenoseal® adhesive, clear | I tube |
| Wood putty |  | I sm. can |
| Sandpaper | 100-, 150-, 220-grit disks | I pk. ea. |
| Steel wool | #0000 | I bag |
| Polyurethane finish | Minwax® clear, water based | I qt. |
| Brush | 4-in. premium bristle | I |

offers a much better price on some of the same basic hardwoods but has a very limited selection of materials.

As you march your take-off sheets around to different suppliers, make an effort to gather more than just a price list. See what you can learn about the supplier—strike up a friendly rapport with the agents, keep an eye out for selection, cull through the stacks, and assess the quality of the material you want for this project and others. As you learn who sells what and who offers the highest quality, the best selection, and the best service, you'll learn where to go for what items.

For example, even though I have that great hardwood dealer I mentioned earlier, I still buy all my hardwood plywood from the large home center. The home center gets its plywood from the same source as the hardwood dealer, but they order such large quantities that they can offer it at a better price. I still buy all my solid-stock hardwood from the local supplier because I get the best selection, quality, and service. While I may pay a little extra for a few solid hardwood pieces that the home center does offer, I usually end up wasting less material, because there are generally fewer defects and because I can get sizes that more closely match what I need. (The home center has a limited selection not only in hardwood species but in board sizes, as well.)

## + SAFETY FIRST

One of the handiest new adhesives is polyurethane glue. Like polyurethane insulation, it expands as it cures, which means it can help fill minor gaps. The bad news is twofold. First, polyurethane glue turns your hands black, so wear latex gloves when you use it. Second, one of the key ingredients is disocynate, a poisonous relative of cyanide, so wash your hands before eating to be doubly sure.

## Choosing Screws

In the United States, wood screws are specified by:

**Style**—Common or bugle head. On a common screw, the thread extends two-thirds the length of the screw. These days, the best option is a bugle-head wood screw, which resembles a drywall screw but has a coarser thread. (See the photo at right.)

**Length**—The length of a wood screw is measured from the tip of the point to the surface of the material into which the screw is driven. So the length of a flat-head wood screw is measured from the tip to the top of the head, whereas the length of a round-head screw is measured from the tip to the bottom of the head.

**Bugle-head wood screw (left); common wood screw (right).**

**Gauge**—Common wood screws were originally manufactured to two different standards. The overall dimensions in the two series were the same, but the number of threads differed slightly. In both series, the size is indicated by a gauge number, from #0 (approximately 1/16 in.) to #30 (approximately 7/16 in.).

**Head style**—Common screws come with a flat, round, or oval-shaped head. Bugle-head screws are flat with a flared (bugle-shaped) shank below the head.

**Drive**—Slotted, Phillips, or square (among others). I prefer using square drives, which are the least likely to cam out.

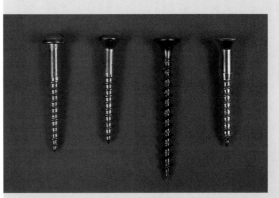

**From left to right: Round head, oval head, bugle head, flat-head wood screw.**

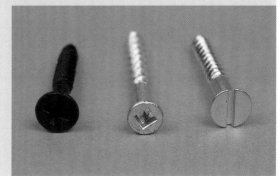

**From left to right: Phillips, square drive, slotted screw.**

# Trim Materials

# CHAPTER TWO

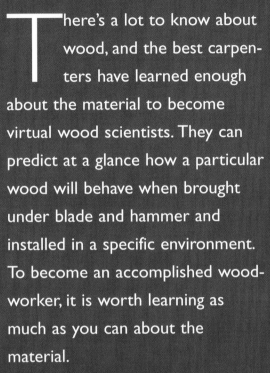

There's a lot to know about wood, and the best carpenters have learned enough about the material to become virtual wood scientists. They can predict at a glance how a particular wood will behave when brought under blade and hammer and installed in a specific environment. To become an accomplished woodworker, it is worth learning as much as you can about the material.

Here are some basic principles you should know about how different woods behave in different conditions. As you sort through piles of lumber and millwork, this information will help you select the right stock for your project. You'll begin to recognize various defects and growth characteristics and understand the consequences they will have on the lasting quality of your work. Become a student of wood. That is, after all, what carpentry is all about.

## IN DETAIL

The terms "hardwood" and "softwood" can be misleading. We commonly refer to trees with needles, such as pine, fir, and hemlock, as softwoods. Trees with leaves, such as maple, ash, and cherry, we call hardwoods. However, this is not always so. Density is the most important predictor of hardness and strength. Dense woods are more difficult to cut, machine, and fasten, as well as less prone to shrinking and swelling. Usually "hardwoods" are denser than "softwoods," but not always.

Southern yellow pine (right) is hard and dense compared to Eastern white pine (left.)

# Choosing Finish Lumber

**When choosing trim stock,** the goal is to select straight, flat, smooth stock that will stay that way. Because wood shrinks and swells with changes in temperature and humidity, a flat, square-edged board tends to grow, shrink, cup, twist, hook, or split as it absorbs water and then dries out. Although they may seem to be fastened in place with screws, glue, and nails, miters can open and close as trim swells and shrinks, making panels warp and cabinet doors stick.

When carpenters talk about a wood's tendency to "move," they're talking about the likelihood of it changing shape, technically referred to as its *dimensional stability.* The dimensional stability of a piece of wood depends on characteristics inherent to that particular species and on characteristics created by sawing the wood into a specific piece of lumber. These characteristics include grain orientation, moisture content, cell structure, and the amount of heartwood vs. sapwood. The presence of defects, such as knots and wane, also affect wood strength.

## Wood characteristics

As trees mature, wood cells form just below the bark during the spring and summer growing seasons. These cells are hollow tubular structures. The thickness of the cell walls increases as new cells are formed during the growing season. During the winter, growth slows down and the cell wall thickness decreases. This change in cell wall thickness from one growing season to the next produces the concentric growth rings we see in a cut log. The strength and stability of any type of wood depends on the thickness of the growth layer, the thickness of individual cell walls, and the properties of the cells. These characteristics vary from species to species and also

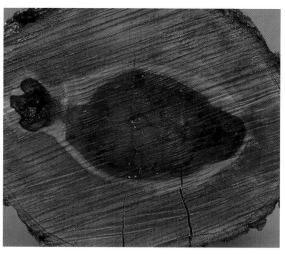

The cell openings that once moved water and nutrients are closed in the denser heartwood (the darker portion of a tree). Hence, lumber cut from this part of the tree is less prone to shrinkage and movement than the surrounding sapwood (the lighter portion).

depend on the site where the tree is growing and the weather during the growing season.

**Heartwood vs. Sapwood.** The strength and stability of wood also depends on the age of the tree. As a tree grows, new cells (called sapwood) are added to the outer regions of the trunk and branches; meanwhile, older sapwood cells gradually change to heartwood.

The difference between sapwood and heartwood is visible in the cross-section of a log. The outer sapwood region is lighter than the darker central heartwood. Both regions provide structural support to the living tree, but water and nutrients flow through the sapwood, as well as the inner bark. The heartwood consists of dead wood cells, and the openings that once moved water and nutrients have closed. That makes heartwood surfaces much more resistant to water movement (and shrinkage) compared to sapwood surfaces.

**Moisture content.** The most important factor that determines a wood's stability is its moisture content (MC). Wood used for interior trim, cabinets, and flooring should be installed at an MC of about 8% to 12%. Wood shrinks as it dries and swells as it absorbs water, so if you install

## Specific Gravity of Common Woods

| Soft Hardwoods | Specific Gravity |
|---|---|
| Black ash | 0.48 |
| Butternut | 0.38 |
| Poplar | 0.40 |
| **Hard Softwoods** | |
| Longleaf pine | 0.60 |
| Larch | 0.52 |
| Douglas fir | 0.50 |

In technical terms, the density of wood is called "specific gravity." Specific gravity is a ratio of the weight of wood fiber, which varies, to the weight of water (1 gram/cubic centimeter).

**Sort materials carefully and cull out the bowed, cupped, twisted, and hooked pieces.**

**Warped Wood**
*These are examples of warp—the distortion of a piece of wood caused by uneven shrinkage.*

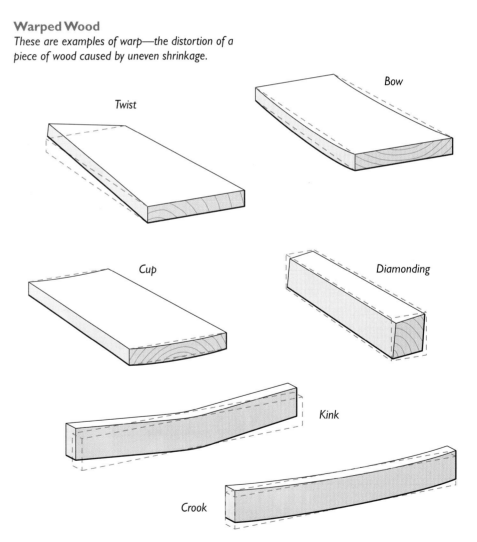

wood with a high moisture content, the joints between two pieces will open up or the pieces will twist, cup, or split as the wood dries to match lower indoor humidity levels. Bear in mind that a piece of wood need not be noticeably wet to have a high moisture content.

Even if you buy straight lumber with an acceptable moisture content, it won't stay that way unless you take care of it. Bring the wood indoors where it will be installed, stack it with stickers between the boards, and let it acclimate to indoor humidity levels. If you store it in a damp basement or garage shop, or if the weather is especially humid, the boards will absorb water from the air and swell.

One of the most reliable ways to ensure an acceptable moisture content is to make sure you buy stock from a reputable dealer. Established lumberyards and hardwood dealers will usually stand behind the material they sell. If you are buying directly from a mill, make sure you talk to someone about the moisture content so that you don't end up with green or uncured stock. Mills

## PRO TIP

*Slow moisture absorption by priming dry paint-grade stock or by sealing the end grain of stain-grade material with an oil-penetrating stain or clear finish.*

### TRADE SECRET

After your stock has acclimated to the environment in which it will be used, make every effort to stabilize the relative humidity level in the building. Run house exhaust fans frequently. In damp climates and during rainy seasons, keep the house windows and doors closed and run a dehumidifier.

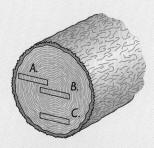

A. Vertical grain
B, C. Flat grain

**Lumber Orientation**
*The grain pattern is determined by the place at which the board is sawn from the log.*

### IN DETAIL

Wood is more stable in the radial direction (the board face lies parallel to the radius of the log) than in the tangential direction (the board face lies perpendicular to the radius of the log).

usually have a moisture meter, so ask someone to verify the moisture content. It should be no higher than 20%; less than 15% is even better. (In drier regions of the country, wood often has an MC of 12% or less.)

Wood rarely dries evenly. This imbalance of moisture content can destroy the wood, which commonly distorts, or warps, as it dries. Because of the way wood cells operate, wood tends to warp in specific ways. Often it *bows* (warps along the length), *hooks* (warps along the width), or *twists* (warps along the width and the depth). *Cupping* occurs when a board becomes uneven across the width due to unequal shrinkage between the board's two faces. Cupping is more pronounced in flat-grain than in edge-grain boards.

## Lumber characteristics

Not all the wood coming from a single log is of the same quality. A lot depends on where the lumber is cut from the log. Wood changes shape least along its length (the dimension parallel to the grain) and most along its width and depth. Which dimension changes depends on the orientation of the grain.

**Grain orientation.** The grain orientation is determined by the place at which the board is sawn from the log. Because most boards are cut

Often warping occurs as individual boards dry in the lumberyard, so it's important to examine them carefully by sighting along their edges. Your eye can be a very accurate gauge of straight lines and square corners.

square from a circularly growing tree, the grain lies somewhere between a completely flat-grain board and an edge-grain (or vertical-grain) board. On most lumber, the grain is oriented somewhere between these two extremes.

**Knots.** Knots form where branches grow off the main trunk of a tree. Knots formed by living branches are usually intergrown with the surrounding trunk, remaining an integral part of the wood. These are called *tight knots*. Knots from dead

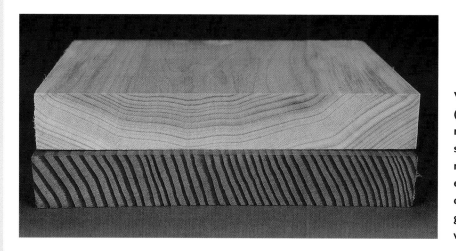

Vertical-grain material (bottom) tends to be much stronger and more stable than flat-sawn material (top). In general, the closer the grain of a board is to vertical grain, the more stable it will be.

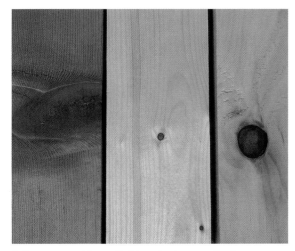

A loose round knot (left) and a large spike knot (right) are unstable and should be cut out of a board. Tight round pin knots (center) are acceptable for paint-grade material.

branches often become *encased* in later trunk growth and are usually found in the central portion of a large trunk. These knots (sometimes referred to as *black knots*) are often loose when the trunk is sawn; if they fall out, they become knotholes.

All knots vary in size and shape, depending on the orientation of the boards cut from the trunk. A knot split radially by the saw extends across the face of the board and is called a *spike knot*. Avoid boards with spike knots that extend across all or most of the face, as these boards will likely break at that point.

When the board is flat sawn, knots appear as rounds or ovals and are called *round knots*. If they are smaller than ¼ in. in diameter, they are called *pin knots.* The form, size, and amount of knots on a given piece of lumber—and whether they are intergrown or loose—are all taken into account when lumber is graded.

## Lumber grades

A single sawn log yields lumber of many different qualities. These lumber qualities are sorted, or *graded,* so buyers can select the type best suited to their needs. Grading is done by visually inspecting lumber at the sawmill. Softwood used for

# Paint Grade vs. Stain Grade

Softwoods and hardwoods represent two distinct lumber categories. Each comes from entirely different types of forests with entirely different harvesting, grading, and distribution systems. Although carpenters constantly refer to these categories, they more often make a distinction between paint-grade and stain-grade materials. These two categories represent types of wood and the precision level of the accompanying work. Paint-grade work relies on a painter to make it all look good. Joints can be caulked and blemishes filled. Stain-grade work, on the other hand, receives a clear finish, so the joints must be tight and the carpenter must pay close attention to the grain pattern and surface texture.

### Common Paint-Grade Options

Choosing paint-grade wood does not necessarily mean choosing knotty boards. Knots show through even the most finely painted surfaces, unless you spend a lot of extra time filling, sanding, and sealing the knots.

- Idaho and Eastern white pine
- Poplar
- Finger-jointed stock
- MDF board stock

### Common Stain-Grade Options

Stain-grade woods focus attention on the *figure* in wood. Figure includes not only the grain patterns but also the

textures and appearances that different wood species and different cuts can take.

### Closed-grain woods

- Clear pine
- Southern yellow (prominent grain)
- Eastern white (uniform appearance)
- Vertical-grain fir
- Maple and birch

### Open-grain woods

- Red and white oak
- Ash
- Mahogany

Examples of tight-grain woods. From left to right: clear pine, maple, and birch.

Examples of open-grain woods. From left to right: red oak, mahogany, and ash.

**PRO TIP**

*Know your lumber grades so you can choose the right stock for the job. The best all-around hardwood grade for the price is No. 1 Common.*

### TRADE SECRET

One of the best ways to control wood movement and prevent splitting, warping, and shrinking is to coat all surfaces with either primer paint or stain before installation. Coating the back surface is called back priming or back sealing. If you're rushed to start, prime only the back. In bathrooms, sunrooms, and pool areas—places where the woodwork may be exposed to extreme fluctuations in humidity, it's also a good idea to prime the ends of any unglued joints before installation.

### IN DETAIL

Most standard trim stock (such as casing and one-piece baseboard) comes with a hollow cut on the back. This practice is called "backing out." Its primary purpose is to weaken the stock, making it less likely to cup when nailed securely in place. The hollow also allows the trim to span lumps in the wall while remaining tight at the edges.

**Elaborate figure in wood can include ray fleck (left) and bird's-eye grain (right).**

trimwork is graded by appearance only, not for strength. Graders look for the shape, size and frequency of knots; checks (splits along the grain, mostly at the ends of boards, caused by uneven drying); pitch pockets (discreet concentrations of resin in softwoods), shake (separation of the growth rings, usually near the pith); and stain (discoloration often caused by fungi).

In other words, grading finish lumber is all about identifying appearance defects. In contrast, structural grades select for strength characteristics, such as grain deviation, density, and growth rings per inch. Therefore, softwood and hardwood finish lumber have separate grading systems.

**Softwood lumber grades.** In general, softwoods fall into *select* and *common* grades. These grades are then divided into categories. Simply put, higher, or select, grades have fewer defects; lower, or common, grades have more.

**Hardwood lumber grades.** Most hardwood is delivered to a millwork shop rough-sawn into general thicknesses, which are designated as 4/4 (about 1 in.), 5/4 (about 1¼ in.), 6/4 (about 1½ in.), 8/4 (about 2 in.), and so on. Typically, hardwood is sold in this rough form in random lengths and widths and usually in a green condition.

Grading is done when the lumber is in this rough condition. The most common hardwood-grading rules used in the United States are those established by the National Hardwood Lumber Association$^{SM}$ (NHLA). These rules specify the percentage of the total board surface (called *surface measure,* or SM) that can be cut out in the form of rectangular, clear-faced pieces (called *cuttings*). Here, *clear faced* means free of knots, wane, shake, and checks.

The highest grade commonly sold is firsts and seconds (FAS), which yields 80% to 90% clear material. It is the most reliable option if you are looking for a uniform surface to finish naturally. However, FAS material is also pricey. I typically use it only when I need long lengths of clear material and don't have the option of cutting out, or otherwise avoiding, small defects.

When price is considered, the best all-around grade is No. 1 Common, which yields about 65% clear material. For many jobs, No. 2 and No. 3

**Medium-density fiberboard (MDF)** is a composite material made of fine wood fibers mixed with a urea-formaldehyde–based adhesive, and then compressed under heat and pressure. Available in a variety of trim profiles, MDF is an exceptionally stable material with a super-smooth surface that takes paint well, but freshly cut edges tend to be extremely fragile and chip easily.

Common work well. These lower grades have some defects but generally not as many as softwood materials of the same designation. If I buy at least 10% more material than I need, I usually can cut out the few knots or relegate pieces with knots and other small defects to hallways, closets, and other less visible locations.

# Choosing Millwork

As the name implies, millwork is any lumber milled into specific profiles for different uses. This can include molding pieces with a variety of curves, such as those found on most base caps, coves, and crowns. Or it can be simple square-edged, thin pieces intended for such uses as lattice and door stop. Larger pieces, such as profiled casing and one-piece base stock, also fall into the millwork category. These days, however, not all millwork is made from wood. Many profiled trim pieces are made from composite wood materials, such as medium-density fiberboard (MDF), or

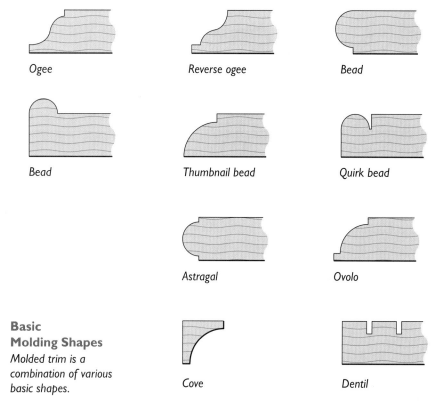

**Basic Molding Shapes**
*Molded trim is a combination of various basic shapes.*

Ogee

Reverse ogee

Bead

Bead

Thumbnail bead

Quirk bead

Astragal

Ovolo

Cove

Dentil

## PRO TIP

*Don't seal the end grain of stock that will be joined with glue. Sealing prevents glue absorption and weakens the glue bond.*

## TRADE SECRET

When making your own trim stock, it's good practice to kerf the back of wide finish lumber. The kerf achieves the same thing as the backing out on milled trim—it helps prevent the board from cupping. I typically run extra-high base-board (made from 1×8 and greater) through the table saw to cut at least three ¼-in.-deep kerf lines in the back. Better still is to run the stock through a dado blade to hollow out the back.

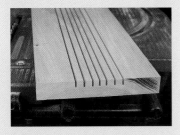

## IN DETAIL

If you're using trim that is backed out or has kerfs cut in the back, keep in mind that exposed ends will reveal the hollow (or the kerf). This means that aprons below windows must have miter returns and butt-joined casing may need a back band.

Custom-cut baseboard (left) has deeper, more ornate profile cuttings than typical stock baseboard (right).

from pure synthetic materials, such as urethane foam and PVC.

### Lumberyard choices

Most lumberyards and home centers carry a selection of millwork, but that selection can be disappointing at times. Generally, a lumberyard buys from one millwork source, and then maybe carries only the most popular profiles from that supplier. As a result, you may be faced with only one choice of profile in each category. But it's worth doing some investigating to discover the various profile choices that may be available. Different yards may carry different profiles, and oftentimes millwork suppliers make available brochures and catalogs of their products. Ask for them, because chances are good that you can find or order a particular profile that you want to use. The trick is identifying which one that is.

Toward that education, you will have to dig around—sometimes deep. Look in old wood-

working and joinery texts. Almost any edition of *Architectural Graphic Standards* will have interesting examples. All of these sources will seed your imagination and interest, develop your eye for trim details and make you more aware of wood-work details in the buildings around you.

### Millwork choices

Although they are expensive, custom moldings are usually your best choice for ornate trim; if you are trying to match a specific period detail, they may be your only choice. For these, you'll need the help of a millwork shop.

The milling machines in a well-equipped mill-work shop have one or more cutterheads. A series of "knives" are custom-cut from steel blanks to the exact dimensions of the molding profile you need; these knives are bolted into the cutterheads. The millwork shop will charge a fixed fee for cutting the knives, so the more trim you order in a particular custom profile, the cheaper it will be per foot. Typically, the cost of a knife is figured by

**Standard Molding Profiles**

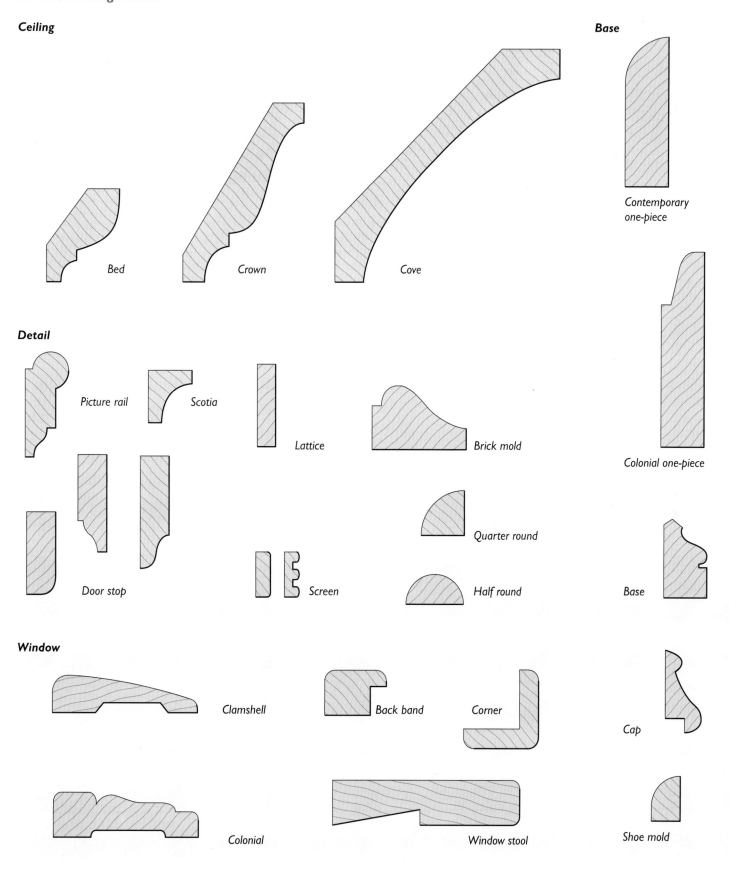

**Ceiling**

Bed

Crown

Cove

**Base**

Contemporary one-piece

Colonial one-piece

Base

**Detail**

Picture rail

Scotia

Lattice

Brick mold

Quarter round

Half round

Door stop

Screen

**Window**

Clamshell

Back band

Corner

Cap

Colonial

Window stool

Shoe mold

## PRO TIP

*If you're trying to match old trimwork, you'll probably need the services of a custom millshop. Ask your hardwood dealer for recommendations.*

## IN DETAIL

The more you work with wood, the more you will want to understand why it behaves the way it does. There's no better guide available than Bruce Hoadley's *Understanding Wood* (The Taunton Press, 2000). Hoadley, a wood scientist at the University of Massachusetts, explains everything from the physical structures and properties of wood to how to cope with moisture and movement and even how to identify different wood species.

## TRADE SECRET

Most mill shops have a huge selection of knives that they have cut for previous jobs. The shop will still charge the fixed "set up" price for bolting the knives into the cutterhead and adjusting the milling machine for your run, but this is much less expensive than cutting new knives, so be sure to look carefully at the existing profiles. A skillful operator can often make a close match with existing knives by setting up the machine to cut only a portion of the profile.

### Period Trim Packages
*Each period architectural style has distinct trim details. Here are examples of built-up chair rail and baseboard profiles representing three common periods.*

**1. Simplified Colonial**

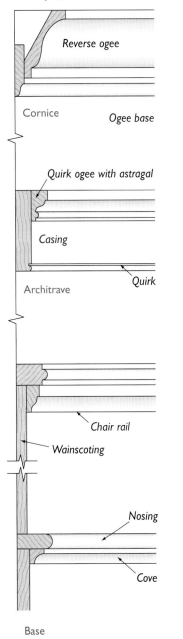

Reverse ogee

Cornice    Ogee base

Quirk ogee with astragal

Casing

Architrave    Quirk

Chair rail

Wainscoting

Nosing

Cove

Base

**2. Federal**

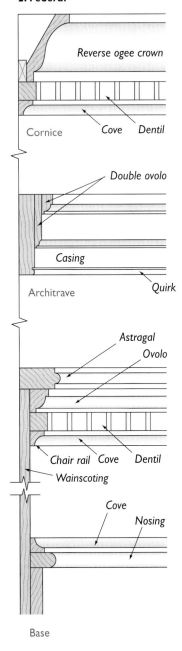

Reverse ogee crown

Cornice    Cove    Dentil

Double ovolo

Casing

Architrave    Quirk

Astragal
Ovolo

Chair rail    Cove    Dentil
Wainscoting

Cove
Nosing

Base

**3. Greek Revival**

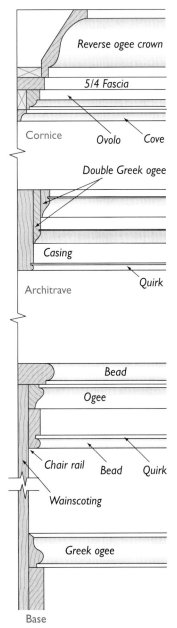

Reverse ogee crown

5/4 Fascia

Cornice    Ovolo    Cove

Double Greek ogee

Casing

Architrave    Quirk

Bead

Ogee

Chair rail    Bead    Quirk

Wainscoting

Greek ogee

Base

## Will the Real Mahogany Please Stand Up?

Order mahogany at your local lumberyard and you'll probably get Meranti (*Shorea* spp.) instead. True American mahogany (*Swietenia* spp.) comes from the West Indies, Mexico, Central America, and South America and is prized for its beautiful dark red appearance, dimensional stability, and decay resistance. A related African mahogany (*Khaya* spp.) is also available, but it is not as durable.

Meranti is much more common than either one of those species. There are four main types of Meranti: dark red, light red, white, and yellow. The dark red is only moderately resistant to rot, and the light red, white, and yellow versions are not durable in exposed conditions. White Meranti dulls cutters because it has a high silica content. The dark red and yellow varieties tend to warp.

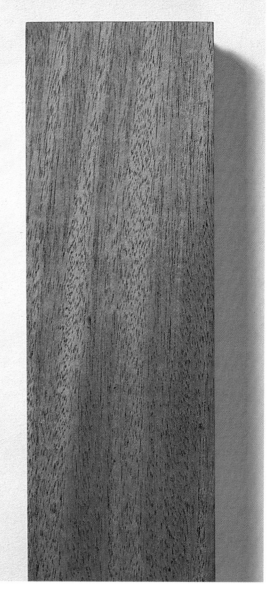

**African mahogany.**

the inch; a set-up charge is usually added on, as well. The final cost of the trim also includes a linear price for the wood.

Millwork that is produced on multiple-head machines costs more, since several knives must be cut for each profile. However, these machines pro-duce a much smoother piece of trim, and the savings in sanding and finishing is usually well worth the extra cost. To order the knives, the millwork shop requires a full-scale drawing of the profile. If you are matching an existing piece of molding, bring a sample, if possible.

# Trimming

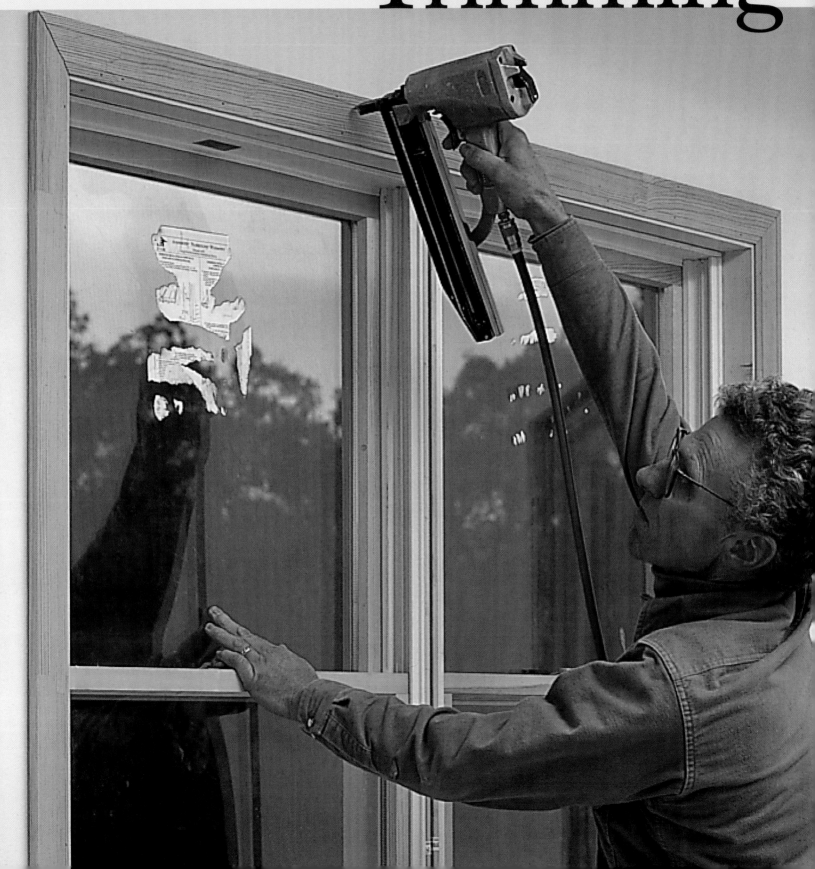

# CHAPTER THREE
# Windows

**M**odern windows are factory-made and come preassembled as a unit that is installed in one piece into the rough opening. Many windows come with a narrow exterior trim, called brick molding, which is prefixed to the frame. And occasionally, a window unit will even include an interior "trim kit." But usually, it's up to the finish carpenter to make and install the interior trim. That's what this chapter is all about.

Of all the trimwork inside a house, window trim is among the most common and conspicuous. Because windows are usually positioned at eye level—in clear sight—the trim you install around the window will be noticed. Just because it is conspicuous doesn't mean it has to be elaborate. Good proportions, capable craftsmanship, and clean detailing can create a simple elegance that will be admired.

**TRADE SECRET**

This is what I carry in my tool belt for all finish carpentry jobs:

- Pencil
- Hammer
- Utility knife
- 25-ft. tape measure
- Speed square
- End nippers (for pulling finish nails)

**IN DETAIL**

When trim turns a corner, such as where a window casing meets a jamb extension, it is usually joined with a reveal. That is, the edges of each piece do not line up but instead are offset about ¼ in. This creates a step, adding an extra shadow line that contributes to the visual complexity of the trim. But a reveal also serves a practical purpose, fooling the eye and hiding any slight misalignments between the meeting pieces. Your eye can easily see when two corners don't meet, but it is unlikely to notice a slight difference in the width of a reveal line.

# Trim Styles

**Trimming a window starts** after the window unit has been installed and the walls have been insulated and covered with drywall. The interior trim covers the gap between the ragged edge of the drywall and the window frame, preferably in a style that matches the woodwork in the rest of the house.

The simplest casing method is to *picture-frame* the window by simply wrapping all four edges of the window with casing and mitering the casing at the corners. This style visually works best for smaller, horizontal windows and for simple trim profiles that are painted to blend with the wall. The other option is commonly called *traditional*

Traditional trim includes casing on the top, or head, of the window (above) and a stool with an apron (below).

window trim. This includes casing on the sides and across the top, or *head,* of the window, while the bottom of the window gets a *stool* with an *apron* below.

The stool often sits higher than the sill to keep wind and water from seeping in below the bottom edge of the window sash. Stock stool profiles have an angled dado that fits into the angled inside edge of the sill. If you are making your own stool, it's wise to mimic this profile to create a weatherproof seal at this critical junction. However, this isn't required with many modern window units, which have a flat inside face to which the stool simply butts.

Keep these points in mind when choosing a window trim style:

- How much attention do you want to draw to the trim?
- What kind of curtain treatments, or blinds, will you use? Wide window stools can sometimes interfere with curtains.
- How square is the window? Picture framing is much more difficult to do when a window unit is wildly out of square.
- How flat is the plane between window and wall? If the window is not exactly parallel to the walls, consider using a reveal joint, bead joint, or corner blocks to mask the discrepancy.
- How much time do you have and how complex a task do you want to take on?

## Anatomy of a window

Before we talk about how to trim a window, it helps to have an understanding of some basic window terminology.

**Jamb extensions.** The frame, or *jamb,* of the window unit may or may not be the same thickness as the wall, and it's not uncommon for the window unit to be narrower. Andersen® window units, for example, always require jamb extensions, because the frames are only 2⅞ in. thick. How-

**Anatomy of a Window**

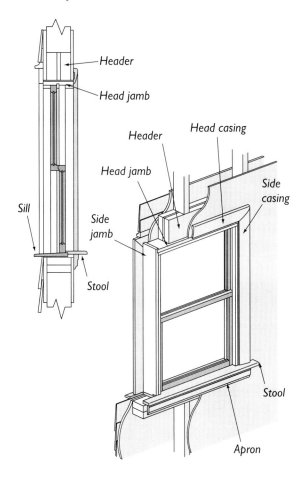

Header
Head jamb
Header
Head casing
Head jamb
Side casing
Sill
Side jamb
Side casing
Stool
Stool
Apron

component of window trim. Window casing covers the gap between the window and the wall and stiffens and supports the jambs.

**Stool.** While we commonly talk about setting something down on the window *sill,* carpenters actually refer to this interior piece as the *stool* to distinguish it from the angled exterior sill. While the sill extends outdoors, past the siding, to shed water away from the building, the stool in contrast is flat and its front edge projects beyond the plane of the interior casing. This edge is often shaped with a roundover or ogee profile, though a square-edged stool is also common. The ends of the stool typically get cut in around the side jambs, creating *horns* for the side casings to sit on. Picture-frame windows don't require a stool; instead, they simply have a bottom jamb extension. Traditionally trimmed windows, however, always have this flat, horizontal piece, which provides a visual foundation for the casing.

**Apron.** The apron sits below the stool. It supports the sill and covers the gap between the dry-

ever, Andersen also provides extensions with a tongue that fits easily and tightly into a groove in the frame to fit standard 4½-in. and 6½-in. walls. Other window companies provide premade jamb extensions, too, so be sure to ask your window dealer about those options when you order your windows. While you may have to pay a little more, I've found that these factory-milled extensions are usually well worth the additional cost. However, if the windows are already ordered, the extensions have been lost, or you are refitting an existing window, you will have to make the jamb extensions yourself.

**Casing.** These are the pieces most commonly referred to as "window trim." The casing is nailed perpendicular to the edges of the jambs and is the most conspicuous—and often the most ornate—

**The ends of the stool are cut in around the side jambs, creating a flat surface for the side casings to sit on.**

## WHAT CAN GO WRONG

A trim carpenter's worst nightmare is a window that was shimmed too tightly when it was installed. This squeezes the jambs on the window unit, causing the sash or casement to bind. Always check the operation of a window before installing the interior trim; otherwise, you will lock the problem into the wall.

## TRADE SECRET

If a window has been shimmed too tightly, it's often possible to fix it by inserting a flat bar and levering out the jamb to free the shim. Then, using a chisel, try to split one shim out. This is often not pretty and can be extremely frustrating. After the top shim is out of the way, knock the jamb tight to the framing, using a wood block so you don't damage the window frame.

### Window Stool Options

**I. Traditional Stool (A)**
The stool extends to the sash.

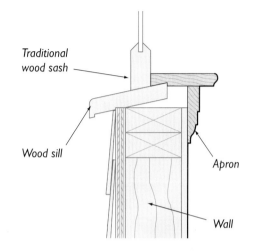

Traditional wood sash

Wood sill

Apron

Wall

**2. Picture-Frame Jamb Extension**
The jamb extension is the same as the head and side jambs.

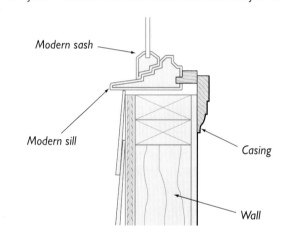

Modern sash

Modern sill

Casing

Wall

**3. Traditional Stool (B)**
The stool extends to the jamb.

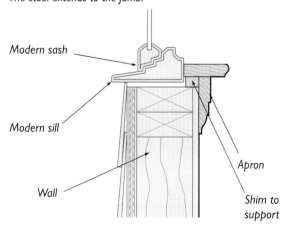

Modern sash

Modern sill

Apron

Wall

Shim to support

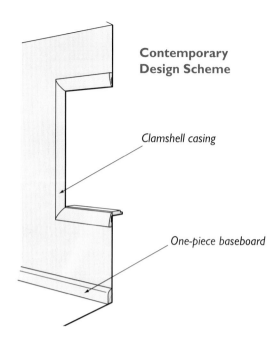

**Contemporary Design Scheme**

Clamshell casing

One-piece baseboard

wall and the window. If a profiled casing is used for the apron, a *return* is usually cut on each end.

## Three standard window designs

There are a wide variety of woodwork design styles from which to choose. However, the majority of my work has always called on variations of just three basic styles.

**Contemporary.** On nearly all of the windows that call for a simple picture frame, I apply a standard contemporary, or clamshell, casing. This material has a thinner inside edge, so it "rolls" easily to conform to window jambs that protrude or sink slightly beyond the wall plane. Basic clamshell trim is widely available in inexpensive Philippine mahogany or white pine. I prefer the pine, since most of the time I want to paint the material so it blends inconspicuously with the wall surfaces.

A contemporary design scheme calls for clean lines and minimal woodwork. If baseboard is used, it may be a simple one-piece wood or rubber cove. Ceiling trim is rarely included.

**Classical.** I use this term loosely to refer to general designs that can be replicated using stock moldings, not as a precise design term to refer to historical restorations. A cornice molding that

# TRADITIONAL HEAD STYLES

Traditional trim lends itself to many variations in design, not only of the molding profile, but also in the way the head joins the sides. These details provide an alternative to simply mitering the corners and serve a practical as well as an aesthetic function. Miter joints—particularly in wider boards—are prone to opening up as the wood dries. The joints below are less apt to open up.

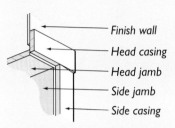

*The head casing and side casing often sit at slightly different planes from one another.*

## Butt Joint

The simplest method of joining the head to the side casings is with a simple butt joint. This style can be cut easily, but any misalignment between the head and the side casings will be noticeable. Also, simple butt joints only work on flat, non-profiled casing stock. If you choose a profiled casing, you must cut a miter or use some other method to join the corners.

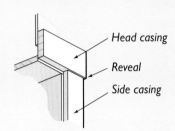

*The head casing projects beyond the face and edge of the side casing to create a reveal.*

## Reveal Joint

If the walls aren't particularly flat or the window isn't exactly parallel to the wall, consider installing a head casing cut from thicker stock (such as 5/4), a detail commonly found in Craftsman-style homes built in the early 1900s. It creates a reveal where the head and sides join, allowing the casing to be slightly misaligned without drawing much attention.

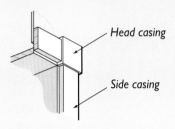

*The block projects beyond both the head casing and the side casing to create a reveal.*

## Corner Blocks

Thicker (typically 5/4) corner blocks inserted between the head and the side casings provide an elegant detail that is common in Victorian buildings. Frequently, these corner blocks have an ornate rosette cut into the center of the square. The rosettes are cut on a drill press with an elaborate bit, or they can be ordered through specialty millwork companies.

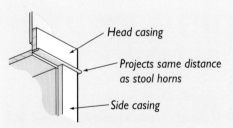

*The bead projects beyond the face and edge of the casings to create a reveal.*

## Bead Joint

Adding a thin molding to the bottom of the head casing creates a bead joint. The molding only needs to be $3/8$ in. to $1/2$ in. thicker than the casing stock, with a very simple profile—usually a roundover.

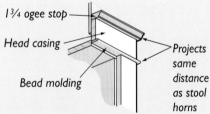

*The bead and ogee stop projects beyond the face and edge of the casings.*

## Capital

For a more elaborate style, add another piece of trim along the top edge of the head casing. Combined with a bead joint, this creates a "capital."

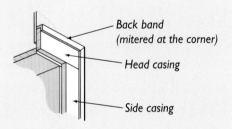

*The side casing butts to the head casing.*

## Back Band

Use with butt-joined or mitered casing to add an extra dimension of visual complexity to the trim. The back band also covers the end grain of butt-joined head casing. It is one of the simplest built-up casing styles.

**PRO TIP**

*Before trimming any window, check its operation. This is your last chance to look for any shimming errors.*

## TRADE SECRET

Trim carpentry requires the utmost accuracy. For the most precise measurements, don't use the hook end of a tape measure. Instead, start your measurements at the 1-in. mark (and be sure to subtract 1 in. from the tape reading). This method is called "cutting the one" or "burning an inch."

## WHAT CAN GO WRONG

If the wall is not flat or the window is installed out of parallel with the wall, the jamb may protrude. The easiest way to fix this is to leave the jamb as it is and skim-coat the wall. Apply the skim coat after the trim has been installed and primed (or the first coat of clear finish has been applied). Use a 12-in. knife to feather the mud over a wide swath.

**Classical Design Scheme**

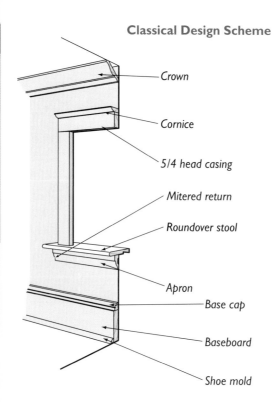

- Crown
- Cornice
- 5/4 head casing
- Mitered return
- Roundover stool
- Apron
- Base cap
- Baseboard
- Shoe mold

**Victorian Design Scheme**

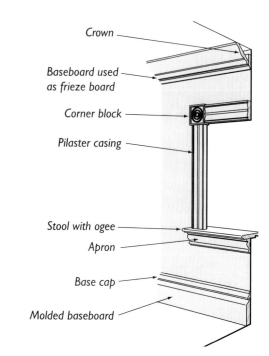

- Crown
- Baseboard used as frieze board
- Corner block
- Pilaster casing
- Stool with ogee
- Apron
- Base cap
- Molded baseboard

wraps around the head casing usually tops classical window and door trim. A reveal joint or bead joint works well to emphasize the look of a "capital." Of course, this style calls for a traditional stool and apron. I typically install an apron made from stock profiled millwork (often called "Colonial" casing), with the ends mitered to a return.

The Classical style blends well into a room scheme that includes tall (1×6 or greater) two-piece baseboards, rectangular plinth blocks at the base of door casings, and crown moldings at the ceiling.

**Victorian.** Trim of this period tends to be much more ornate and fanciful than trim in the more restrained Classical style. Around windows and doors, plinth blocks and corner blocks with rosettes fit well. Corner blocks are most often used with pilaster casings, which are milled in a symmetrical, linear design. The stool typically has a generous ogee cut along the front edge. The apron is wide, often built up with an ornate band

and coped or mitered back at the ends. The high level of ornament carries over into the rest of the woodwork as well. Elaborate crown moldings (often with an added frieze), picture rails, and chair rails with beaded wainscoting are all signature details of the Victorian period.

# Jamb Extensions

The type of jamb extension you need depends on the width of the window and the width of the wall. The simplest type is a piece of square-cut stock that butts against a square window edge. I make this type if I need an extension that is less than 1¼ in. wide. It allows me to nail 8d finish nails through the edge of the piece into the window frame. In this case, I usually leave a reveal (if there's room in the rough opening), offsetting the extension from the existing window jamb.

If the extensions are much wider, however, I need another way to attach them. In that case, I usually rip a "step" in the back face. This allows

## Common Jamb Extension Profiles

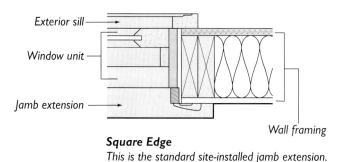

**Square Edge**
*This is the standard site-installed jamb extension.*

**Back Stepped**
*This wide jamb extension (supplied by Eagle Windows) can be made on site.*

**Tongue and Groove**
*This manufacturer-supplied jamb extension is typical of Andersen Windows.*

me to secure the extension with a screw through the step, or shoulder. I always preassemble these wider extensions into an extension "unit," which makes for tighter joints and a faster job.

## Window prep

Before trimming any window, first check to see how well the window operates. If the window has been shimmed too tightly, this is your last chance to correct it easily. It's also your last chance to seal it well against air leaks.

Begin by cleaning off any drywall compound that may have collected at the edges near the rough openings. Cut back any shims that extend past the window frame. Not all windows need jamb extensions. Ideally, the jambs should extend about ¹⁄₁₆ in. beyond the wall surface. Check this by holding a square or straightedge against the wall and sliding it over the window.

In some cases, the existing window jambs may even stand a bit proud, projecting into the room slightly. If this is the case, you may need to plane the window jambs. If the jambs project into the

room more than ¼ in., don't cut them or try to plane the entire amount off. Not only is this difficult to do, but you will also wind up creating a twisted plane, making it nearly impossible for the casing to lie flat. In most cases, you're better off leaving the jambs as they are, installing the casing, and trying to fill the gap between the edge of the

**Before installing jambs, cut back any shims that may have been used to install the window units.**

### IN DETAIL

Ideally, a window should be sealed soon after it is installed. In new construction, this should happen before insulation and drywall. Many builders still stuff the gap between the rough opening and a window or door with fiberglass insulation. However, fiberglass will not stop air. Instead, fill the gap with nonexpanding foam. Make sure you do *not* use expanding foam, which can overexpand, bowing in the jambs and even jamming the sash so the window can't operate.

### IN DETAIL

When the windows and walls are out of whack by more than ³⁄₃₂ in., I scribe the jamb extensions to fit and rip each piece to width with a jigsaw. I rip slightly wider than the mark so there is enough extra to plane off the saw marks; otherwise, these will show at the reveal.

**Before installing casing, check the corners of windows to make sure that the jamb comes out far enough. In this case, the drywall had to be shaved back to allow the casing to fit.**

casing and the wall with a feathered skim coat of drywall compound.

### Measuring jamb extensions

To determine the width of the jamb extensions, hold a straightedge on the wall so that it spans the corner of the window opening. Measure from the straightedge to the window frame and add ¹⁄₁₆ in. This means the jamb will extend ever so slightly into the room, making it easier to apply the casing.

It's common for a window to be slightly out of parallel with the plane of the wall, so measure all four corners with a straightedge. If the measurements differ by more than ³⁄₁₆ in., then I custom-cut the extensions, tapering them slightly to match the true dimensions. If the width of the jamb extensions differs by less than ³⁄₁₆ in., I split the difference and adjust the discrepancy with the

casing. That is, I cut all jamb extensions ³⁄₃₂ in. narrower than the widest piece. This means that at the deepest corner, the jamb extension dips ³⁄₃₂ in.; at the shallowest corner, it protrudes ³⁄₃₂ in. Typically, I can make up for this later by rolling the casing in slightly where the jamb dips and caulking between the casing and the wall where it protrudes.

### Ripping jamb extensions to width

I mill jamb extensions from *S4S* stock (boards that are sanded smooth on all four sides), typically eastern white pine, and use the same material for the window frame. Choose the best grade your budget will allow. I prefer using clear material rather than trying to conceal knots, which are hard to seal and which sometimes loosen or fall out. Because of the way that light from the window falls across the jambs, any surface blemish will likely show up, even through paint.

Always cut your jambs from solid wood, not MDF or plywood. Composite materials tend to split when nailed or screwed through the edge. For narrow, square-edged extension pieces, I try to plan my rips so I have a factory-milled edge facing into the room. For example, if I am ripping 1³⁄₈-in.-wide pieces, I cut them from 1×4 stock. With each pass through the table saw, I hold one factory edge toward the fence. This way, each piece has one clean edge that doesn't require planing away the saw marks, which would otherwise be seen along the reveal.

## Picture-frame assembly

Using a small square or a reveal block, I first mark the edge of the window jamb for the reveal. This provides precise marks from which to measure the length of each extension. If the extensions are less than 1 in. wide, I cut each piece to length, and then nail it directly to the window with 6d to 8d nails, depending on the width of the extension. Nail both sides of each corner and every 10 to 12 in. in between. When nailing, be sure to drive the nails straight (in line with the existing window jamb), not angled toward the window, to avoid blowouts.

If the extensions are wider than 1 in., I find it's always worthwhile to preassemble the four jambs. I lay the pieces out on a piece of plywood or large benchtop and screw the corners together. It is then much easier to nail the entire assembly to the window jamb rather than aligning and installing each piece separately.

## Traditional-trim assembly

If the window has a stool, I cut and install this piece first, as described below. If the extensions are wide, the stool can be built right into the jamb assembly and installed as a unit.

The stool should be installed as level as possible with stable shims below it. If something heavy is set on the stool (or worse, someone stands on it when hanging curtains), the stool must bear considerable weight. The stool can split or sag if it's not well shimmed.

**Cutting a window stool.** The first thing to determine is the width of the stool. To do this, hold a straightedge across the lower corners of the window and measure the distance from the window to the straightedge. But instead of rip-

## Making Wide Extension Jambs

Wider extensions require three passes through a table saw. I first rip the material to width, just as I do for narrow stock. Here again, I try to keep a factory edge facing the room. Then, I rip a step in the back of each piece. Ripping this step, or shoulder, requires two passes through the table saw. The first pass is made with the board on edge and the blade set at an angle (this depends on the depth of the extension piece you are cutting). The height of the blade should be set so it reaches from the outside corner of the board to the inside corner of your second cut.

The second pass cuts the back face. This cut should be made about 1 in. from the edge that will butt against the window or so that it reaches the kerf made by the first pass. Set the height of the blade at about 3/8 in. for 1× stock, so the blade reaches to the middle of the board. Keep the blade vertical.

1. First rip a long bevel.

2. Rip square; set the blade height to reach the end of the first bevel.

3. Completed "step" cut.

PRO TIP

*Make sure a window stool is well shimmed below. Sometimes it must bear considerable weight.*

## IN DETAIL

Not all miters are a perfect 45-degrees. Use this geometrical layout for odd angles or to create a miter for an out-of-square opening.

**Bisecting a Miter**

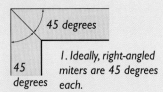

*1. Ideally, right-angled miters are 45 degrees each.*

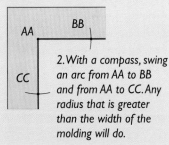

*2. With a compass, swing an arc from AA to BB and from AA to CC. Any radius that is greater than the width of the molding will do.*

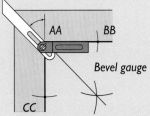

*3. Next, swing arcs from BB and CC to create an X. The line drawn from this X to AA and through the joint bisects the angle. Transfer it with a bevel gauge and set the chopsaw.*

**Predrill the ends of the boards before screwing together the prebuilt jamb unit.**

ping to this width, you need to add the thickness of the casing, plus another ¾ in. This means the stool will protrude beyond the plane of the casing by ¾ inch. This added dimension may be larger if a wider window stool is preferred. I think a ¾-in. projection looks best. If you are using a back band or building up casing from several overlapping pieces of material, make sure you account for the added thickness when figuring the width of the stool. When you know the overall width, rip the stool out of a piece of 5/4 stock.

Next, cut the stool to length. The total length of the stool must account for the width of the window (including jamb extension reveals, if necessary), the casing reveals on both sides, the width of the casing on both sides, plus an additional ¾ in. beyond each casing end. I usually rip the edge that will butt to the window with a very slight 2-degree bevel. This back-cut ensures that the stool meets tightly against the window.

## Making a Window Stool

1. Measure and rip to width: Extension width, plus casing thickness, plus ¾-in. projection.

2. Measure and cut to length: Inside window width (including jamb extension reveals), plus twice the casing reveal, plus twice the casing width, plus twice the ¾-in. projection.

3. Mark the centerlines of the window and the stool stock. Then align the stool over the window opening, holding the stool against the wall.

4. Check for parallel between the stool and the window.

5. Set the width of the scribe tool to the distance between the window and the edge of the stool stock, then scribe the horns.

6. Hold a square along the inside of the rough opening and mark the inside corners of the horns.

7. Cut the horns.

8. Mill the outside edge of the stool with a router and a ½-in. roundover bit.

9. Sand the edges; if the trim will be painted, prime both sides of the stool.

I was taught to cut a miter return on the end of the horn of window stools, but these days I rarely do it. Instead, I simply round over the end grain with a router. However, on some open-grain hardwoods, such as oak and mahogany, the end grain may be visually quite different. In that case, it's best to cut a miter return. Be sure to leave the

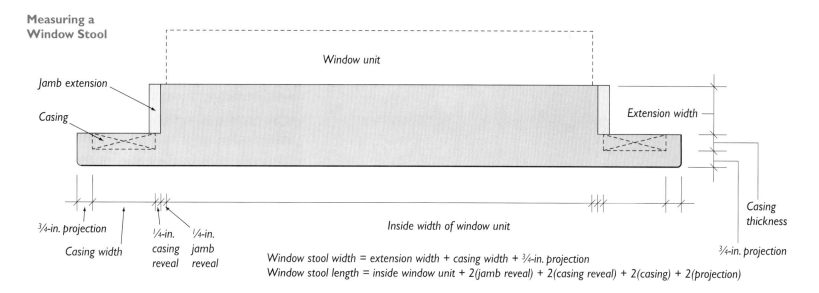

**Measuring a Window Stool**

Window unit

Jamb extension

Casing

Extension width

Casing thickness

¾-in. projection

Casing width

¼-in. casing reveal

¼-in. jamb reveal

Inside width of window unit

¾-in. projection

Window stool width = extension width + casing width + ¾-in. projection
Window stool length = inside window unit + 2(jamb reveal) + 2(casing reveal) + 2(casing) + 2(projection)

stool horns long, so that you have plenty of stock to cut the return blocks with a matching grain.

After cutting the sill stock to length and width, round over the outside edge and the ends using a router. I typically use a ½-in. roundover for 5/4 stock. Depending on the trim style, you may wish to cut a different profile. Victorian trim, for example, often calls for an ogee edge profile. A more contemporary design, on the other hand, may call for simply easing the edges slightly with a block plane. After shaping the edges, smooth them with sandpaper. If the trim will be painted, prime both sides of the stool.

Now you are ready to cut the horns on the end of the stool. I start by marking the center of the window and the center of the stool piece that has been cut to length. Then, I draw a centerline across the bottom of the rough opening, so that I can align the stool to this mark, and hold it against the wall in its exact location. Measure the distance between the window and the stool piece on both sides of the window to be sure the stool and the window are parallel. If the window is slightly out of parallel with the wall, shim one end of the stool stock. Spread a scribe tool (such as a compass) to the distance between the window and the edge of the stool, and then scribe the ends.

When creating an ornate apron built from overlay moldings over flat casing stock, be sure to cut the stool wide enough to accommodate the extra thickness.

**+ SAFETY FIRST**

Table saws are extremely prone to kickback. To minimize this, feed material slowly and evenly. Don't force it. Keep your body out of the line of travel of the blade. Use extreme caution when cutting short pieces.

## PRO TIP

*Make sure you don't shim too much. It's rather easy to bow the sill upward when pushing wedged shims together.*

## TRADE SECRET

I own several handplanes, but my all-time favorite is the Stanley® 96 low-angle block plane. This small plane has a blade set at just 12½ degrees from horizontal. This low angle allows the blade to shave the material without a lot of chatter. While specifically designed for shaving through dense end grain, I use it for almost all the hand-planing I do, whether it's back-cutting a miter or erasing saw marks from the edge of ripped stock.

## IN DETAIL

Garnet sandpaper gives a smoother finish than most other abrasives because it is self-sharpening—the grit fractures as it abrades, offering new, sharp edges. It also stays much cooler than other sandpapers do, making it a good choice for power sanding when other abrasives might scorch the wood surface.

**After aligning the center marks of the stool and window, the notch can be marked.**

**Then use a square placed along the inside of the window opening to mark the end cuts.**

Finally, hold a square along the inside of the window opening to mark the end cuts. I make the cuts for the horns with a miter saw and finish the cut in the corner with a short handsaw. Once the stool is cut, I test-fit it. This may require scribing, both against the window and along the horns. When it fits tight to both the window and the wall, I build it into my extension assembly.

**Installing a jamb/sill assembly.** Before installing the assembly in the window opening, I drill pilot holes through the shoulder of the stepped extension stock and at an angle through the bottom of the stool. These pilot holes keep the jamb extensions and stool from splitting when the screws are driven home. The pilot holes also allow me to get the right angle, so that the screw doesn't burst through the inside of the window frame.

Install the extension assembly with 1½-in.-long screws. I use a long Phillips bit in a magnetic holder to reach between the assembly and the drywall. This can be a pretty tight squeeze, and sometimes even the magnetic bit holder is too wide. You may want to use a hex extension bit

from a screw gun, which is easier to maneuver in tight spaces. One screw near the corners and every 16 in. in between is sufficient to hold the assembly securely in place.

Be sure to secure the sill, too. Begin by shimming under it. Typically, the gap between a finish sill and the rough framing is quite wide, so I usually cut strips of plywood to use as shims. Stack these and slip a couple of new shims on top, and then snugly wedge the entire stack into the opening. Just make sure you don't shim *too* much. It's rather easy to bow the sill upward when pushing wedged shims together. Depending on the width of the window, a stack of shims should be located in one or two places in the middle, as well as below each side jamb (though sometimes the side jambs of a window unit extend hard to the rough sill, so this isn't always necessary).

Nail through the sill with 8d to 10d finish nails. As a last step, nail through the face of the horns into the trimmer studs below with a 10d nail in each side. Be sure to predrill the holes before nailing to prevent splitting the horn.

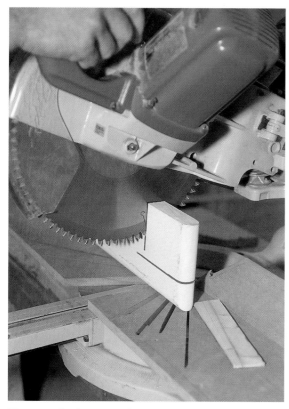

First, cut the horns to size on the window stool.

Then test-fit the stool.

The stool may need to be scribed to fit tightly against a large window unit.

A jamb with a step ripped into it can be installed with screws.

43

## PRO TIP

When buying a nail set, look for an ASME B107.49M-1998 designation on the package. This indicates the set has been tempered for striking.

## IN DETAIL

Before filling nail holes, set the nail head just 1/16 in. below the wood surface—enough to hold a daub of wood putty. Before filling the hole, I prime the woodwork. Then, taking a daub of painter's putty on the end of a 1-in. putty knife, I fill the hole and use the knife to scrape. After the putty has dried, I sand the filled holes lightly with 100-grit sandpaper before finish painting.

## TRADE SECRET

When nailing near the end of a board, blunt the nail's tip to prevent it from splitting the wood. This can be done quickly by holding the nail upside down and striking the tip with a few blows of a hammer.

# Picture-Framing a Window

A picture-frame window has four pieces of casing, each mitered at 45-degrees in the corners. The challenge here is to get all four miters tight. This depends on having a perfectly square window.

One way to check a window for square is by holding a framing square inside the jamb at each corner. However, a better way is to measure and compare the diagonals across the window. If the two diagonals are unequal, the window is out of square and you should try to correct it, if possible, by forcing in new shims around the window. If you can't do that, you may be able to compensate by making adjustments to the reveal width and the miters. Correcting for an out-of-square window requires patience and experience. Just remember that all the miters are connected; an adjustment to one will affect all the others.

## Layout

The casing lines up on the edge of the window jamb (or jamb extension) about 1/4 in. in from the inside corner. This 1/4 in. is the reveal. Begin your layout by using a small tri-square to measure and mark the jambs for a consistent reveal, as follows:

**1.** Adjust the blade to a short 1/4 in.

**2.** Hold a pencil at the end of the blade. Some carpenters file a notch at the end to hold the pencil, but it's fine just to hold the pencil tight to the end of the blade.

**3.** Move the square and pencil down the length of the window jamb, scribing a line parallel to the edge.

These marks tell you where to nail the casing. You don't have to make a continuous line. The most important thing is to mark at the corners and somewhere in the middle of each length. In

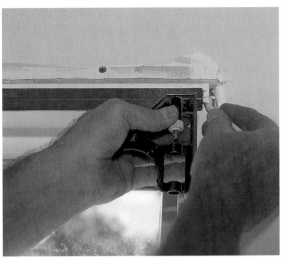

**A tri-square provides a ready guide for marking the edges of a window jamb with a consistent reveal.**

fact, I have made the mistake of marking a continuous line for the reveal on unpainted trim, and then regretted it. After the casing was nailed on, I could see my line in most places, and it proved very difficult to erase.

## Measuring length

Begin by holding a long piece of casing stock across the head of the window, lining it up along the edge of the jamb on the reveal marks. Make a mark on the casing at the reveal mark in each corner. This marks the inside, or *short point,* of the miter. (The outside corner of the miter is called the *long point.*) Make a slash across the casing to indicate the direction of miter cut from short point to long point, so that you don't lose track when you make the cut.

On a production job, you would repeat this procedure for the other three sides before installing the pieces. However, if the windows aren't perfectly square, or if you anticipate other problems, start by just laying out and cutting the two angles on the head casing first. Then tack it in place and continue measuring and cutting the remaining pieces one at a time. This way, if the miters are wildly off, you can hold rough-cut lengths directly over the tacked-up head casing

**A quick way to calculate the head piece for mitered casing is to measure the distance between the outside edges of the side pieces; this is also the same length as the apron.**

and mark the actual short and long points of each angle. Doing this takes patience and skill, but if you keep the casing long, you can often sneak up on it.

## Cutting miter joints

If you're new to cutting miters, a picture-frame window is a good place to practice. You can use a wood miter box or a chopsaw. I usually use a chopsaw, which is faster and makes it easier to precisely adjust an angle if a window frame is not perfectly square.

I cut the two miters on the head piece first, and then tack this piece in place (without sinking the nails home). Then, I cut matching side pieces and tack them in place, securing them at the top ends only. Last, I cut the bottom piece. Before making final miter cuts, measure between the long points on the two side pieces, and then double-check the short-point measurements.

## Constructing a Wood Miter Box

1. Make a basic box using 1×4 hardwood for the sides and a scrap of 2×6 framing lumber for the bottom. Screw the sides to the bottom, making sure all the parts are flat, square, and parallel.

2. Find the center and mark across the top of both sides.

3. Measure the inside width of the box. Make a mark on both sides of the center-line the same distance as the width of the box. (Your marks should make a square.)

4. With a square and a pencil, draw three cutlines down from the marks; the center will be for square cuts and the other two for 45-degree cuts. Draw these inside the box and then extend them to the outside.

5. Tack a piece of scrap 1×2 alongside one of the angled lines, as shown. Use this to guide a handsaw as you cut three slots—one at the centerline and one at each of the 45-degree angles.

6. Use the slots to make square and mitered cuts.

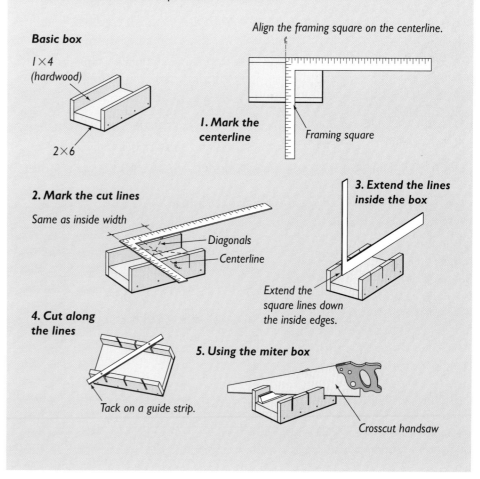

**Basic box**

1×4
(hardwood)

2×6

**1. Mark the centerline**

Align the framing square on the centerline.

Framing square

**2. Mark the cut lines**

Same as inside width

Diagonals

Centerline

**3. Extend the lines inside the box**

Extend the square lines down the inside edges.

**4. Cut along the lines**

Tack on a guide strip.

**5. Using the miter box**

Crosscut handsaw

## PRO TIP

*Use a biscuit in the butt joints on casings. A biscuit helps align the joint, making it stronger and less likely to open up over time.*

## TOOLS AND MATERIALS

In finish carpentry, a pneumatic brad nailer doesn't just speed the job along—it also improves quality. Not having to bang on the trim with a hammer means that there's less chance of shifting the alignment of the pieces.

## WHAT CAN GO WRONG

Butt joints may not fit perfectly if the window jamb isn't perfectly flush with the surface of the wall or if the wall or the trim material is warped. For example, if the jamb protrudes slightly, the head casing may tip back slightly, opening a gap at the joint. Try either back-cutting the end of each side casing or shimming the head casing slightly forward.

## Nailing

When you know all the pieces are the right length and the miters are cut at the right angles, you can finish driving your tack nails home. For standard clamshell casing, I use 4d (1½-in.-long) nails along the thinner inside edges into the window jamb and larger 6d (2-in.-long) nails along the thicker edges into the trimmer studs beneath the drywall. I typically pair nails at about 45-degrees along the trim, mirroring the pair near the miters. I finish by driving a 4d nail along the outside edge of the casings near the long points of each miter. Make sure you set the nails with a nail set and fill the holes afterward.

# Traditional Casing

Casing for a traditional stool-and-apron window always has a few less miters than a picture-frame window; sometimes, it has no miters at all. Nevertheless, when you factor in the extra time involved in cutting and fitting the stool, traditional trim takes more time to install.

## Mitered casing

One of the simplest traditional-trim styles uses standard Colonial casing stock mitered at the head. Begin as if you were picture framing the window: First, measure the length of the head piece, then cut the two ends of the head piece and tack it in place. Next, cut the two side pieces. I typically cut the meeting side of the miter first, leaving the sides long. The horns on the sill get in the way of fitting the miter, but you can usually bow the length in toward the wall to get the miter in the right location. Once I know the angle is right, I cut the side piece to length by measuring from the stool horn to the long points of the head casing on each side.

## Butt joints

Generally, when I fit traditional stool-and-apron trim, I don't cut any miters at all. I typically use "flat stock"—standard 1×4 or wider finish lumber—and join the head casing to the side casings with a butt joint. In this case, the side casings butt square against the bottom edge of the head casing.

The biggest advantage of butt joints is that if the wood dries and shrinks, the joint can be tightened back up. Miters, on the other hand, can't be tightened up later because the wood shrinks unevenly across the joint, changing the miter angle (remember that wood shrinks at different rates lengthwise and crosswise to the grain).

**Traditional Casings**

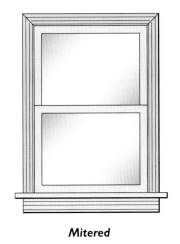

*Mitered*

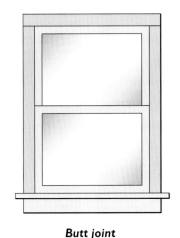

*Butt joint*

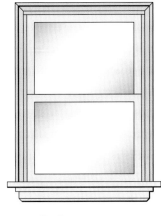

*Built-up casing*

**Measure the length of side casings from the top of the stool to the reveal mark across the head jamb.**

## Fixing Bad Miters

### The Angles Are Off

If the opening is not perfectly square, a miter can be open at the heel or at the point. Correct this by recutting the angle on a chopsaw. The adjustment may only be half a degree or so, and it may be difficult to set the saw at this angle. Instead, consider using wedges between the casing and the saw table. Remember: Recutting the angle will change the length of your material, so always check the window for square. Cut some test miters until you get the angles right.

### The Vertical Is Off

A miter can also be open if the wall is not perfectly flat. To tighten up the joint, "back-cut" the miter slightly. Very slight gaps can be shaved with a low-angle block plane or filed down with a rasp. More extreme gaps can be back-cut with a chopsaw by propping the casing up on the table.

### The Surface Plane Is Off

If the wall is not flat and the window is cocked, the top surfaces of the two pieces may not meet perfectly even though the gap is tight. In this case, carefully shave down the casing with a chisel or shim one piece slightly forward. If you shim the casing forward, caulk the gap before painting.

When using butt joints, I usually start with the side casing. First, measure the length of each side, resting the end of the tape on the stool horn, and find the length to the reveal mark for the head casing. After cutting the side pieces to length, tack them in place. Then measure the distance across the top of the window, from the outside of one side to the outside of the other. This is the length of the head casing.

I usually cut the head casing square at this dimension. However, some carpenters prefer to let the head casing overhang the sides by a small amount—usually the same amount at which the horns project, or about ¾ in. In this case, add 1½ in. to the length of the head piece (or simply use the same measurement as that of the stool) before cutting square.

Once the head casing fits, nail it in place along the jamb edge first, then along the top edge. Space the nails 10 to 12 in. apart and slope the nails at a slight downward angle to drive the head piece tightly into the butt joints. Later, if the framing and trim shrink and the butt joint opens up, you can tighten the joint by hammering against a block.

## PRO TIP

*Don't use alcohol-based putty on shellacked surfaces. The alcohol will dissolve the finish. Instead, use a color-matched wax pencil.*

### TRADE SECRET

Instead of using a hammer claw to pull out finish nails, I usually use a pair of end nippers. These bite easily into the nail shank at any point. Whenever possible, I try to pull the nail out from the back. The finish head pulls right through the material, resulting in a clean finish surface. This works especially well when disassembling woodwork that will be reinstalled.

### IN DETAIL

When filling nail holes in unfinished wood, it's often difficult to get a good match because filler absorbs stain and finish differently than wood does. Instead, I use Color Putty®. First apply any stain and all finish coats, allowing them to dry before applying the Color Putty. Wipe off any excess with a paper towel dampened with denatured alcohol. Wear a pair of latex gloves to prevent the pigment from staining your fingers.

The length of the head casing equals the distance between reveal marks on the side jambs, plus twice the width of the casing stock.

Nail through the stool to secure the top edge of the apron.

## Fitting an apron

The apron should be cut from the same width stock as the casing. However, it doesn't need to be cut from the same material. In fact, it's common for the apron to be cut from profiled stock, even when the casing is cut from flat stock. This is a customary Classical detail. (See the drawing on p. 36.) Find the length of the apron by measuring the distance from one outside edge of the casing to the other. This measurement should be the same length as the head piece, if the head doesn't project past the side pieces.

If the apron is cut from flat stock, simply center the apron below the window stool and nail it in place with 8d nails every 10 in. If the apron is wide, you may only have nailing along the top edge. I usually plane (with a block plane) the top edge of the apron so that it fits tightly to the stool. I like to clamp the edge with a Quik-clamp® to draw the stool and apron together, then drive a few 8d finish nails to secure them.

If the apron is cut from profiled stock, the ends must be visually "closed" and you will have to cut miter returns. In that case, the length of the apron runs from long point to long point, with very short triangular pieces of stock meeting at the miters and returning to the wall.

## Built-up casings

As the name implies, many pieces of casing can be matched and layered to build up extra-wide, very ornate trim profiles. Use your imagination to combine stock moldings, but just keep in mind that every piece you add doubles the installation time.

Built-up casing typically includes a "ground" made of flat stock, with a variety of trim pieces layered over the ground to create complexity. Make sure you account for the added dimensions in both thickness and width before ripping your stool stock. I always cut short pieces of all the

**A simple overlay molding easily dresses up standard casing.**

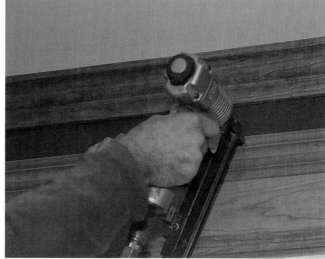

**Overlay moldings can be applied with 4d finish nails. It doesn't take much to secure them to flat casing stock; gluing is unnecessary.**

layers and make a model to figure out how they will fit together. Then I measure the built-up dimensions directly off the model.

Built-up moldings can be simple bands applied over flat stock to create an ornate capital or a more elaborate layered casing. An example of a more elaborate built-up casing includes a ground of S4S 1×4, which is installed with butt joints like a conventional casing. After cutting, fitting, and installing the ground (holding it to the reveal line on the window jamb), a secondary molding cut from 1¼-in. by ⅜-in. ogee door stop is layered on top. The ogee aligns to the outside edge of the ground, laps over the butt joints, and is mitered across the head. The built-up casing is now 1⅛ in. thick.

After cutting and installing the overlay molding, a cove molding is installed around the outside edge of the ground. This ⅞-in. piece covers up the lap joint between the ground and the overlaying molding. Once it is installed, the total width of each side casing is 4⅜ in.

Use 4d finish nails to secure overlay pieces to the ground. There's no need to drive through the ground into the trimmer studs. This is a good place to use a pneumatic brad nailer, if you have one.

Unless your ground lies perfectly flat, you may see significant gaps between the overlay moldings

and the underlying trim. Butt joints may need to be belt-sanded if the head and side pieces of the ground are significantly out of alignment. However, if the trim will be painted, a judicious bead of caulk goes a long way toward hiding minor defects.

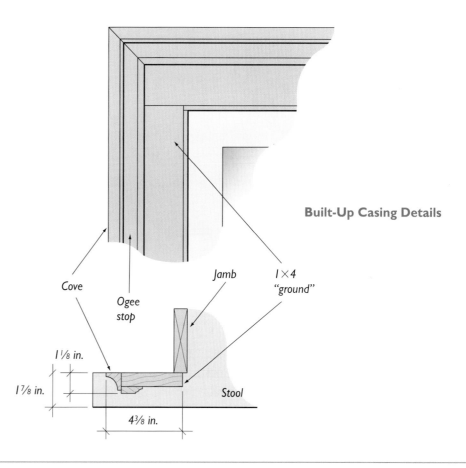

**Built-Up Casing Details**

Cove

Ogee stop

Jamb

1×4 "ground"

1⅛ in.

1⅞ in.

Stool

4⅜ in.

# Running

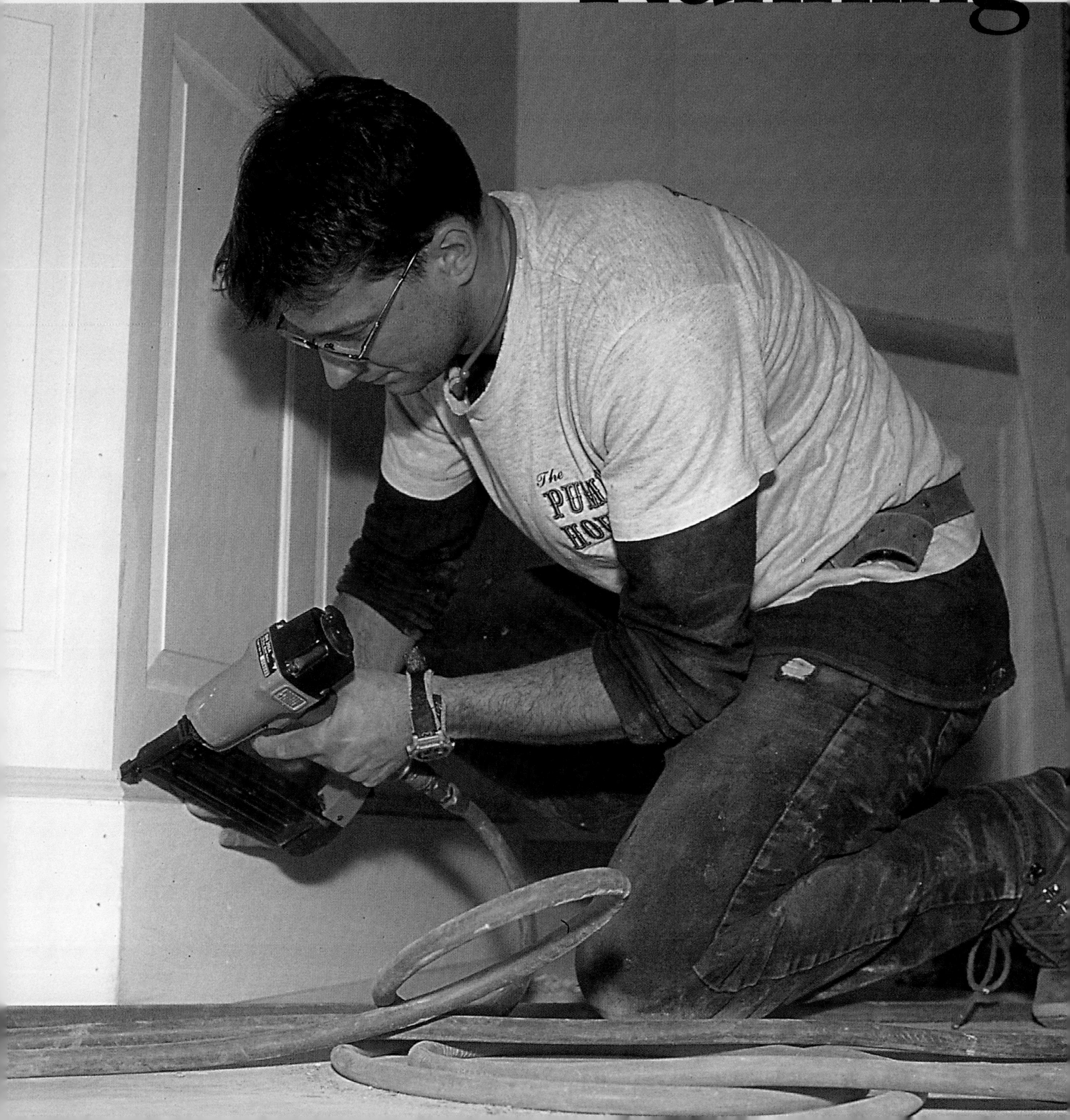

# CHAPTER FOUR

# Baseboard

When it comes time to install baseboard, old-time carpenters gladly yield to younger members of the crew—the ones quickest at getting up and down from the floor and the ones least likely to complain about crawling around the job site on hands and knees.

Baseboard—also called *mopboard, skirting,* or simply *base*—should be one of the last pieces of trim to install. After the door casings have been installed, the heating vents and electrical outlets roughed in, the drywall taped, and the floors sanded, baseboard ties it all together. As the name suggests, it provides a visual base for the vertical, or "standing," trim, such as door casing and wainscoting. But it's much more than a decorative band. Baseboard hides the inevitable gap between the wall and floor, and it covers the all-important space around the perimeter of a wood floor that allows floorboards to expand and contract.

### TRADE SECRET

If a door casing or plinth block is out of square, you may be able to scribe the cut. Most of the time, however, the length of baseboard fits into a corner, so you don't have any room to run the baseboard long while scribing. In this case, you have two options: a bevel square or a "preacher block." The bevel square works best if you can hold it on a piece of scrap baseboard instead of on the floor. If you have several of these cuts to make, you can save time by making a preacher block. Run the baseboard long past the door casing, then slip this U-shaped block over the baseboard and trace the exact line of the cut.

# Baseboard Layout

**O**ne-piece baseboard can be run quite quickly, but it isn't very forgiving of sags and dips in the floor and walls. If you're working in an old house, plan at the very least to run an additional shoe molding to conform to discrepancies in the floor. Thinner one-piece base stock will bend enough to conform to an uneven wall surface. But if you're running a tall, traditional base, plan on using a separate base cap as well. Regardless of the material you use, work out the how the base will intersect with the door casing well in advance. These are details that you want to work out when you're installing the door casing, not running the baseboard.

Begin the baseboard layout by defining the areas in the house where the baseboard will be most visible, and then sort through your material. Cull through the pile, tagging the best-looking lengths for the most conspicuous areas. The entry, living, and dining areas typically receive the best material, while hallways and bedrooms receive less-than-perfect lengths. The worst material—pieces with large knots or slightly twisted, bowed, or crooked lengths—can be cut into smaller pieces for use in closets and behind radiators.

## Installation order

Running baseboard efficiently starts with a room survey to determine a logical order for installing it. Look at all the corners and note the inside and outside ones. All outside corners will be mitered, so these should be done last. All inside corners of profiled base should be coped. (Coping is not

### Baseboard Types

*Wood baseboard comes in a wide variety of styles and sizes. Choose a design best suited to your particular situation.*

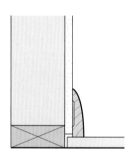

**Small Base (or Sanitary)**
*The baseboard equivalent to clamshell: This small rounded profile is thin enough to bend, conforming easily to walls, and on long runs even bowing to fill small dips in the floor.*

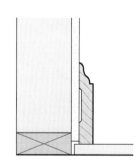

**Profiled One-Piece Base**
*This can add elegance to an interior trim package but is not very forgiving of wavy walls and sagging floors. Save this option for use in new construction, when you know the walls will be reasonably straight and flat and the floors level.*

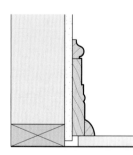

**Tall Base with Base Shoe**
*In older houses, opt for a three-piece baseboard. While the main piece stays rigid, the flexible cap conforms to waves in the wall and the thin shoe mold bends to match sags in the floor.*

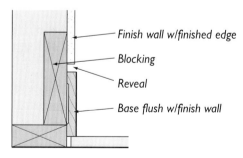

— Finish wall w/finished edge
— Blocking
— Reveal
— Base flush w/finish wall

**Flush Base**
*This looks simple but is deceivingly difficult to install. The backer must be continuous, the framing plates must be even, and the bottom edge of the drywall must be level and finished with J-bead.*

# EFFICIENT BASEBOARD LAYOUT

To understand efficient baseboard layout, think in these terms:

- **Lap piece:** cut square, lapping behind the coped piece.
- **Meeting piece:** usually coped, overlays the lap piece.

Start on the wall opposite the main entrance to a room, and then work from each end around the other sides of the room, as shown in the drawing. Arrange the cuts so that the joints are not in a direct line of sight. Work from long lengths to shorter lengths.

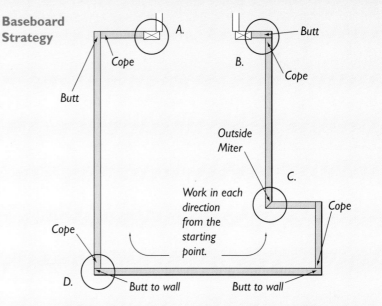

**Baseboard Strategy**

Cope · Butt · A.

Butt · B. · Cope

Butt

Outside Miter

C.

Work in each direction from the starting point.

Cope

Cope

D. · Butt to wall · Butt to wall

*Start baseboard installation on longest wall opposite entry*

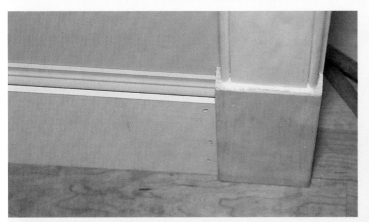

**A. Base to door casing.**

**B. Inside corner at door casing.**

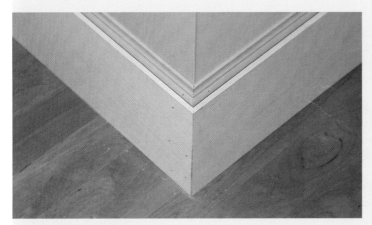

**C. Outside corner.**

**D. Inside corner.**

PRO TIP

*To save time, measure as many runs as possible. Cut those pieces all at once and then install them.*

## TRADE SECRET

The key to efficient baseboard installation:

**Measure**

- Survey the room and identify which pieces to install first.
- Premeasure all lengths, beginning with the lap pieces that have square cuts on both ends.
- Mark profiles in the corners with scrap. Measure the length for the meeting pieces.
- Lay out the outside corners on the floor.

**Cut**

- Chop the square ends, back-cutting one or two degrees.
- Cut the reverse bevels for coped inside corners.
- Cope the profiles, then fit them.
- Cut the outside miters.

**Install**

- Place lap pieces first, pinning them in place along the bottom wall plate only.
- Fit the coped joints and secure the meeting pieces.
- Measure and cut the scarf joints for any spliced runs. Glue and nail the joints.
- Drive finish nails along the bottom wall plates.
- Nail higher up on the baseboard into each stud.
- Set the nails as you sink them.

**In a butt joint for baseboard, both pieces are cut square. The lap piece (left) is installed first, and the meeting piece (right) butts into it.**

necessary for square-edged baseboard.) While you could miter the inside corners of profiled base to save time, the results are sure to be disappointing, because corners are almost never perfectly square. Coped joints are much more forgiving. (For more on coped joints, see p. 56.)

Start on the wall opposite the main entrance to a room, and then work from each end around each side of the room. The goal is to arrange the cuts so that the joints are out of your direct line of sight. If possible, avoid looking "into" a joint.

I call the first piece installed the lap piece. This length is cut square on both ends and laps behind the butt (or coped) ends of the meeting pieces. Ideally, you want to start on one of the longest walls with this first piece. Since both ends are cut square, you're less likely to spoil a long piece of stock by cutting a bad cope.

### Accurate measurements

Going back and forth from your saw to where you are installing baseboard will slow you down, plus your back and knees will be a lot happier if you don't have to get up and down too frequently. You can cut down on a lot of this back and forth

if you first measure as many runs as possible, cut those pieces all at once, and then install them.

Measure lap pieces first, bending the tape into the corner. For long runs, cut the board 1/16 in. long so that it can be sprung into place. Don't cut it too long, however, or you may crack the drywall tape in the corner when the piece is installed.

To premeasure the meeting pieces, use a scrap of baseboard, hold it in the corner, and trace the profile onto the wall (see the left photo on the facing page). Then measure to the pencil line of the profile (see the right photo on the facing page). This will be the short point for the cope on the meeting piece. Take care during this initial measuring phase to check the tilt of the baseboard. If the wall is out of plumb or there is a heavy buildup of drywall mud at the base of a corner joint, the baseboard will tip back. Unless you take this into account, you'll end up with an ugly gap in the corners.

To avoid creating corner gaps, I make it a habit to quickly check for square when measuring with scrap pieces. Hold a Speed Rip Square™ against the floor. If the base is more than 1/8 in. out of plumb, make a note to cut the meeting piece at a

**If a heavy buildup of drywall mud prevents the baseboard from fitting snugly in the corner, the first piece of baseboard in a corner (called the lap piece) can be undercut, as shown. This way, only the top edge—the only part of the end that will show—contacts the wall.**

To accurately premeasure baseboard, use a scrap piece of stock, and mark the front edge of the lap piece.

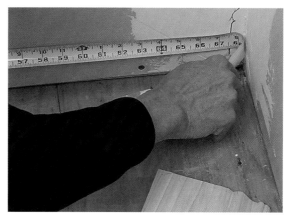

Measure to line. On profiled base, this will be the short point of the reverse bevel for a cope cut.

slight angle to match. If it is close to square but not exact, go ahead and cut the meeting piece square, and then push the joint tight with a flat bar, as described below (see Nailing, p. 60).

There will always be a few pieces to measure after you have some base pieces installed. One common example is a length that is coped on one end and precisely butted into the door casing on the other. In this case, cut the cope first, leaving the end long, and then come back later with a preacher block to mark the cutline for the butt end.

# Cutting Baseboard

Although it would seem logical that most baseboard is simply cut square or at 45-degree angles, I find that this is rarely the case. More often, cuts vary to conform to the irregularities of the walls and the particular type of baseboard stock you're using. Most of these cuts can be made easily by hand with a power circular saw and a simple coping saw. But I've found that a power miter saw speeds the task and vastly improves the quality of my cuts.

## Butt joints

In theory, butt ends are cut square. But to allow for some relief from the mud buildup in corners, I always back-cut the butt ends of my lap pieces slightly by about 1 to 2 degrees. I often undercut the lap ends, too, making a slight angle in the

board to allow room for mud buildup at the bottom of the corner. Remember: The lap piece only needs to fit tightly into the corner at the top; the bottom will be covered by the coped joint.

I rarely back-cut the meeting ends—those that butt the lap pieces (or coped pieces, if the baseboard has a profile). However, I often overcut or undercut the piece if the lap piece is tipped in or out. Before making your final cut, double-check the corner with a Speed Rip Square on the floor. If the wall is tipped so that the baseboard pieces do not fit square when held flat against the wall, measure the difference off square and cut the lap piece at a slight angle to match.

## Coping inside corners

Unless the trim style calls for corner blocks, I always cope inside corners when the baseboard has a profile. Coping has two clear advantages over an inside miter. First, a coped joint is very forgiving of

**PRO TIP**

*Coping a molding with a profile has many advantages over an inside miter, especially when it comes to shrinkage and expansion with seasonal changes.*

## TRADE SECRET

In any kind of carpentry work, one thing that will slow you down is constantly changing your saw settings. Try to avoid this by planning ahead and grouping the different types of cuts together. For example, first cut all the back-cut butt joints with a 1-degree or 2-degree angle going in one direction. Next, cut the ones going in the other direction. Then cut all the reverse miters for your cope joints.

## TRADE SECRET

I use a pneumatic trim nailer to quickly nail off baseboard. This can be speeded along even more by running a tape measure on the floor and holding a Speed Rip Square along the top edge of the baseboard, as shown in the photo. The tape shows you exactly where the stud layout falls, and the Speed Rip Square gives you an instant plumb line along which to nail.

out-of-square corners and irregular wall surfaces, so it's easy to cut a tight joint at the outset. Second, a cope stays tight over time. Changes in humidity cause wood to shrink and swell across the grain. So even if a miter fits perfectly when installed, it will tend to open and close with the seasons. A coped joint, on the other hand, can only open if the lap piece shrinks in thickness (perpendicular to the grain) or if the meeting piece shrinks lengthwise (along the grain). In both cases, this type of shrinkage is minimal.

**1.** Reverse the bevel. Start with a reverse-bevel cut—a 45-degree bevel with the long point on the back of the board. Overcut the length for this bevel by a healthy ¹⁄₁₆ in., so that the piece will be slightly long and the leading edge of the coped profile will bite into the piece of wood it meets. If you know the lap piece is tipped in or

out, scribe the angle first, and then cut the bevel to that line (a compound angle).

**2.** Mark the edge. The reverse bevel defines the profile you will remove; however, the actual joint is really closer to a butt joint. Therefore, I use a small tri-square to mark the square that will butt into the piece of baseboard lapped behind. I may still back cut beyond this square line, but it gives me a reference at the edge.

**3.** Trace the profile. With the flat side of a pencil, blacken the line of the profile. This helps define it, making it much easier to follow while you are sawing the wood.

**4.** Rough out with a jigsaw. A coping saw is a rather delicate instrument, and I find it slow and painstaking to use for the long, straight run on one-piece base stock. Instead, I use a jigsaw to

### Block Corners
*Corner blocks on both the inside and the outside aren't affected by shrinkage or differential movement and offer an alternative to cope and miter joints when running baseboard.*

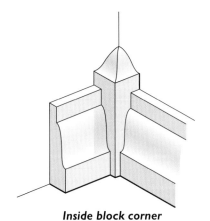

*Inside block corner*          *Outside block corner*

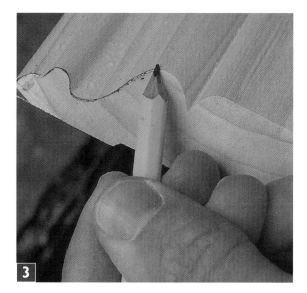

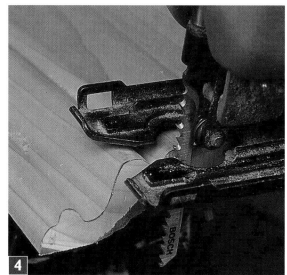

### Base Meets Door

*Well before you start running baseboard, you must work out the details of where it will intersect the door casing. Above all, make sure the baseboard does not project farther from the wall than the door casing.*

*Flush*

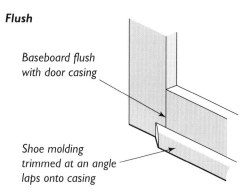

Baseboard flush with door casing

Shoe molding trimmed at an angle laps onto casing

*Reveal*

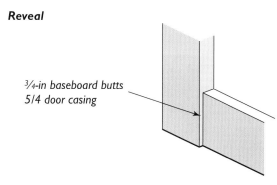

¾-in baseboard butts 5/4 door casing

*Plinth block*

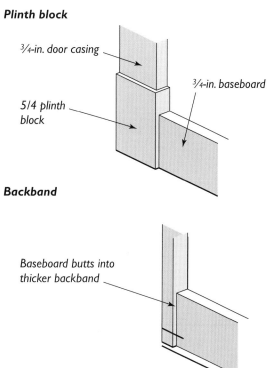

¾-in. door casing

5/4 plinth block

¾-in. baseboard

*Backband*

Baseboard butts into thicker backband

rough out the joint. Some carpenters cut the entire joint with a jigsaw, and then fine-tune it.

**5.** Cope. I switch to a coping saw to navigate the twists and turns of the profiled parts of the base stock. Don't force the blade through tight curves. A coping saw works best when it's used delicately. In tight curves, I break the cut into sections, coming at the blackened line from different directions, gradually sneaking up on it until I have a smooth curve. To fine-tune the cut, don't be shy about employing a host of other tools, such as a rat-tail rasp, a piece of sandpaper, and even a utility knife, to carve and smooth the cut.

PRO **TIP**

*When coping, break the cut into sections, coming at the blackened line from different directions, gradually sneaking up on it until you have a smooth edge.*

## WHAT CAN GO WRONG

The most common mistake people make when measuring cope moldings is butting the tape to the outside of the lap piece. Always measure a coped molding from the *back* of the lap piece.

## IN DETAIL

Occasionally, it's necessary to terminate baseboard at the end of a run. In that case, cut miter returns on the ends of both the baseboard and the base cap. Take care when cutting these tiny miter returns on a power miter box, as the small piece you want tends to fly up into the blade guard or, worse, straight at you. I find it works best to cut the miter first, and then adjust the saw for a straight cut. Glue and miter the return piece to a matching miter on the run of the baseboard or base cap.

**5**

**6**

**6.** Fit. Getting a tight fit may take a few tries. As you can see here, the lap piece with which the cope mates essentially passes through the coped profile. So when fine-tuning the joint, you must be sure to clear all the material behind the leading edge of the profile. I call the coped edge a "knife-edge" to emphasize that the cope has been back-cut. Because I overcut the length by a good $\frac{1}{16}$ in. at the offset, the knife-edge actually bites into its mate when the baseboard is snapped into place.

Coping well takes some practice, so be patient. It's particularly challenging with hardwood, but with persistence you can get a crisp edge that mates precisely. Soft pine cuts easily—sometimes too easily. The resulting edge may end up a little ragged, but if the baseboard is cut slightly long, the joint will still squeeze together tightly.

## Outside corners of rooms

Outside corners of rooms rarely end up being perfect 90-degree angles. Even though the corner may be perfectly framed, metal corner bead and the joint compound covering it conspire to make the angle either fatter or leaner. I find that most of the time the outside miter on each run of baseboard must be cut at an angle slightly greater than 45 degrees.

To find the exact angle, I draw the corner right on the floor using scrap baseboard stock as a template. If I'm working on a finished floor, I lay down wide masking tape on the floor and make marks on that.

Take a short scrap of base material and hold it against the wall, letting it run past the corner. Draw on the floor along the bottom edge to mark the outside face of the baseboard (see photo 1 on the facing page). Do this on both sides of the corner, so that the two lines intersect (see photo 2 on the facing page). Now take your length of baseboard stock and cut or cope the opposite end so that you can hold it tightly in place, with the end running long, past the outside corner. Mark the face of the board where the floor marks intersect (see photo 3 on the facing page), and then mark

# MAKING AN OUTSIDE CORNER

**1.** Position the rough length along one side of the outside corner and draw a line where the base meets the floor near the outside corner. Also, mark where the base stock hits the wall corner on the top edge of the base stock.

**2.** Position the other piece and mark the floor near the outside corner. The result is two crossing lines on the floor. Also, mark the wall corner on the top edge.

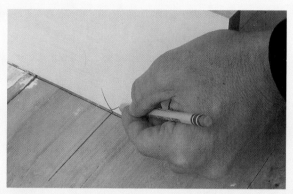

**3.** Reposition the first piece over the crossed lines and mark the intersection on the base stock.

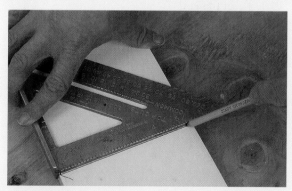

**4.** Using a Speed Rip Square, draw a line on the base stock to define the long point of the outside miter.

**5.** Using a bevel square, draw a miter from the outside corner line (long point of miter) to the mark on the top edge that indicates the wall corner (short point of miter).

*Since cutting baseboard often requires some trial and error to fit the joints, it's always best to cut a little long and sneak up to the fit.*

## WHAT CAN GO WRONG

Shoe molding is the best solution for hiding gaps at the bottom of baseboard. However, with one-piece base, there's no way around it—you have to scribe. In this case, tack the baseboard to the wall so that it's level and scribe the bottom edge. If the floor is way out of level, you may have to cheat, holding the baseboard out of level so it doesn't end up too short on one side of the room. The easiest method of scribing to the floor is with a carpenter's pencil. Hold the flat edge of the pencil on the floor, keeping the pencil square to the baseboard.

the top edge of the board where it touches the corner of the wall. Since corner bead is slightly rounded on the corner, give yourself some room on this second mark—at least a fat ¹⁄₁₆ in. from where the back of the baseboard touches the wall.

The two marks you have made on the baseboard define the limits of your outside bevel cut, but you have a couple of options on how to cut it.

- If you're reasonably sure that the corner is plumb, draw a square line across the face of the baseboard from the floor mark, and then connect the line on the top edge with the wall mark. This defines your bevel, which may or may not be 45 degrees (see photo 4, p. 59).
- Find the angle by laying a bevel square over your floor marks, and then use them to set the bevel angle of your saw (see photo 5, p. 59).
- If the corner is reasonably square but not plumb, hold your board on the table of the miter saw. Lift the board slightly as you eyeball the two marks until they line up with the blade. Hold the board at this elevation and shim it up on the miter saw table to hold it there, and then cut.

Chances are good that your outside corner is neither plumb nor square, and so you will probably want to try some variation of these options, such as drawing the square line across the face of the baseboard, tipping it ever so slightly off plumb, eyeballing it, and then cutting it. If the top edge is covered by base cap, you can also cheat by cutting the baseboards slightly long, so that you are out from the bum corner. If you're far enough away, the baseboard fits when cut at a perfect square and bevel. A little experience (meaning lots of trial and error) will help you find the right solution.

## Splicing base runs

Baseboard stock typically comes in 12-ft. to16-ft. lengths, so it's not difficult to get full-length pieces along most walls. However, in a long hallway (or if you run short on material) it's not uncommon to splice pieces in the middle of a run. For this, you'll want to cut a scarf joint—sometimes called a lapping miter—rather than butt sections together. A scarf joint stays closed better than a simple butt joint should the two lengths of material shrink slightly (see the sidebar on the facing page).

If you're running tall baseboard and have to splice sections together, be sure to break the joint over a stud. This will give you solid backing when you nail the scarf joint together. And before cutting the scarf joint, cut and fit any coped joints or outside miters on the other end first. This allows you a certain margin for error in case you have to recut the joints.

# Installing Baseboard

Ideally, you've cut most of the baseboard to length, so now you're ready to start installing it. However, don't start nailing baseboard until you have at least two pieces that meet, so that you can check the fit. First tap one piece into place, using a block of wood to protect the baseboard from the hammer, and check the fit against the floor. If you're not using shoe molding, you may need to scribe the bottom edge to conform to an uneven floor. Double-check the fit of the joint as well before nailing it home. There's a lot of trial and error involved when fitting baseboard, so it's always best to cut a little long and sneak up on the fit.

## Nailing

When you're satisfied with the fit, nail the baseboard in place. I start nailing in the middle of a run, leaving the ends loose until I have the meeting pieces in place. This gives me some flexibility if I have to shim the ends for a tighter fit in the corners.

I typically use 2½-in.-long (8d) finish nails for nailing most ⅝-in.- and ¾-in.-thick base stock to walls. For thicker 5/4 stock, I use longer 10d nails. When I'm running low, one-piece baseboard (less than 4 in. tall), I place two nails—one low into the bottom wall plate and another higher into each stud. When I'm running taller, three-piece baseboard, I nail only high (in the top third of the baseboard) into each stud, and then let the shoe molding hold the bottom edge. This helps prevent tall baseboard from cracking, which is likely if it is pinned high and low by a pair of nails.

Unless you're using a nailer, predrill holes when nailing near the end of a piece to avoid splitting the wood. I don't worry about splitting when I use a nailer, which shoots a thinner, blunt nail. The blunt end actually pushes a plug of wood in front of it, while a hand-driven finish nail has a thicker pointed end that wedges itself through the grain, splitting the fibers apart as it's driven inward. Hardwood stock is more prone to splitting. Softwood fibers crush together to accommodate a nail, but denser hardwood fibers are pushed aside, increasing the chances that the grain will split all the way through.

When predrilling outside miters, don't drill the full length of the nail. Just drill through the first piece and only a short way into the second, so

## + SAFETY FIRST

As their name implies, air nailers are serious tools—and as dangerous as any guns regulated under the Brady Law. Make sure you understand how your nailer works, always wear safety glasses, and follow through with regular maintenance on both the nailer and the compressor.

## Cutting a Scarf Joint

To splice two pieces of baseboard, cut a lapping miter. The joint should break over a stud.

**Cut the first miter on one length of baseboard, position it on the wall, and measure the matching piece to the short point of the first piece.**

**1.** Cut the ends of each piece square with a 45-degree bevel. The long point of the bevel on the first piece should be on the back of the board, the long point for the second piece on the face of the board.

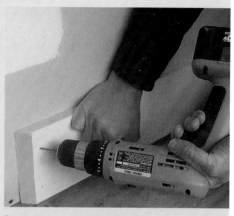

**Predrill for nails and fasten the first piece to the wall.**

**2.** Install the first piece with the long point against the wall. Angle the nails so that they are parallel to the open face of the bevel.

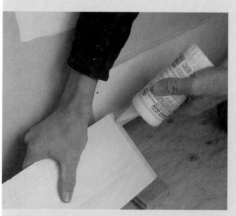

**Glue the miter end on the second piece.**

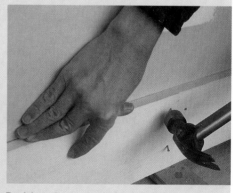

**Position the second piece, predrill, and nail it home.**

**3.** Glue the open bevel, and then install the second piece over it. Using 4d (1-in.) nails, nail directly through the joint perpendicular to the first set. If you don't have a nailer, predrill for the nails. The joint tends to slide around with glue on it, but the predrilled nails make it easier to position the second piece in its exact location.

## PRO TIP

*If you're using a hammer to install trim, always predrill for the nails at the end of a piece to avoid splitting.*

## TRADE SECRET

To predrill for nails, clip the head off a finish nail with a pair of end nippers and chuck the nail into a drill. Use the same size nail for boring the perfect size hole. Only the cuts near the nail points do any work—the rest is friction—so this predrill bit will get pretty hot. In hardwoods, it may even smoke a little. Change the nail/bit frequently to keep it sharp.

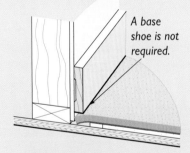

*A base shoe is not required.*

## TRADE SECRET

Carpet is often installed after the baseboard. In this situation, use ⅜-in. or ½-in. spacer blocks when nailing on the base, so that it will be held in the right position. If you have to install baseboard before installing wood or some other flooring material, make sure you at least have a sample of the flooring. Use the sample as a spacer so the flooring can be slid underneath the trim later.

Predrill before nailing at the outside corners to avoid splitting the ends.

Nail from both sides of the outside corner and set the nails.

To avoid blowouts when using a pneumatic trim nailer, align the nailer in a straight line, sighting along the baseboard.

Install the beveled length to the wall, breaking over a stud. Measure for the second piece to the short point of the miter on the first piece.

Cut the bevel on the second piece to length, glue the mitered end, and position it over the first piece.

After nailing through the scarf joint, smooth the joint with sandpaper.

## Nailing Three-Piece Baseboard

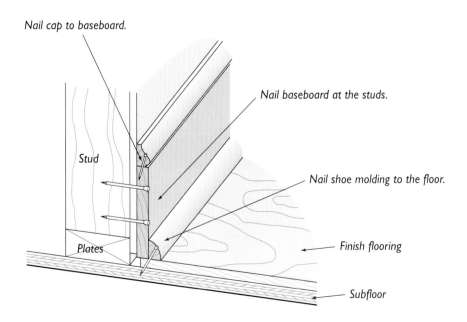

Nail cap to baseboard.

Nail baseboard at the studs.

Stud

Nail shoe molding to the floor.

Plates

Finish flooring

Subfloor

## WHAT CAN GO WRONG

Don't attach a single piece of trim to both a wood floor and a masonry wall, as the differential movement between the dissimilar materials will loosen the connection over time or cause the wood to split.

## TRADE SECRET

After cutting the reverse bevel for a cope, make two straight chops in a power miter saw to clear away the bulk of the material, as shown. This makes it easier to maneuver the coping saw around the turns of a tight profile.

that the combined length of the hole equals about three-fourths of the nail's total length. This ensures that the nail will grab tightly.

When nailing outside miters, I use 4d nails for ⅝-in. stock, and 6d for ¾-in. stock. I typically use four nails—two from one side and two from the other—to lock the joint in place. Tall base (over 6 in.) may call for three nails on each side. Small ranch base generally needs only two nails—one on each side.

Inside corners need to be nailed only near the corner—two nails in the lap piece and two in the meeting piece. I angle these nails toward the corner to draw the joint together, but the nails are not intended to pass through both pieces of base, only into the wall framing.

As mentioned previously, base can roll back at the corners. Hopefully this is taken into account when measuring and cutting the run. When nailing, I always begin in the center of the run, leaving the ends free so I have the option at the corners to tip the top of the baseboard forward (a flat bar helps here) so the lap piece butts tightly with a square-cut meeting piece. A few thin shims behind the corner may be needed to press the joint closed. If slight, a visible gap between the wall and the back of the baseboard can be caulked (or hidden by base cap). However, if the gap is too wide, the meeting piece may need to be recut at an angle to compensate.

## Base cap and shoe molding

Obviously, base cap and shoe molding must be run after the baseboard is nailed in place. Again, inside corners should be coped and outside corners mitered. If there is a long run that requires a splice, be sure to stagger it so that it doesn't align with a splice in the baseboard. First, install a premitered length to the wall, breaking it over a stud. Then measure the second piece from the short point of the bevel on the first piece and cut it to

length. After double-checking the fit, put some glue on the joint, nail through the scarf into the stud, and smooth the joint with sandpaper.

Shoe molding should be nailed only to the floor. Base cap should normally be nailed to the top edge of the baseboard. A 4d nail usually works for both. Do not glue base cap or shoe molding to the baseboard. Each piece should be free to expand and contract at its own rate.

## Details

If we only had to deal with outside and inside corners, running baseboard would be a snap. But often there are plenty of niggling details that slow down the job.

**Meeting stairs.** The connection between the stair stringer and the base trim is often one of the most difficult details to work out. Ideally, the baseboard and stringer materials should match in thickness and in height. The shoe molding can

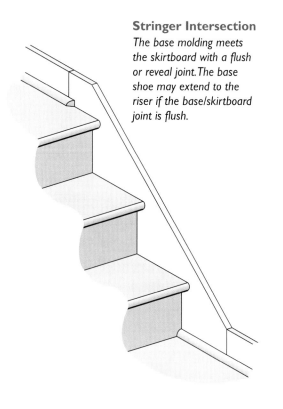

**Stringer Intersection**
*The base molding meets the skirtboard with a flush or reveal joint. The base shoe may extend to the riser if the base/skirtboard joint is flush.*

## CUTTING AN OUTLET

Locating and cutting an outlet often requires painstaking precision.

**2.** Find a metal electrical box that is the same size as the one to be installed in the wall and position it on the transfer centerlines. Use the box to mark the exact dimensions for the cut.

**3.** Use two drill bits. The first is a ⅜-in.-diameter twist bit to bore for the tabs. This allows the outlet to be placed in the metal box after the baseboard and box have been installed. The second is a larger spade bit to drill a starter hole in one corner of the cutout.

**1.** Cut the baseboard to length and lay it face down in front of the outlet. Using a tape, transfer the distance from one wall to the centerline of each box, measuring first on the wall and then on the back of the baseboard. Also, measure from the floor to the bottom edge of the box, and then transfer that distance onto the baseboard, measuring from the bottom edge.

**4.** Insert the blade through the large starter hole and cut out the outlet opening with a jigsaw.

**5.** Nail the completed baseboard into place.

## IN DETAIL

When tall baseboard meets a built-in cabinet, the kick space often falls below the top of the baseboard. While this can sometimes be an awkward detail, Tom Morris Construction of Burlington, Vermont, handles it nicely here. The baseboard has been notched under the kick space and the base cap dies into a wide cabinet face frame.

## TRADE SECRET

In retrofit work, it's sometimes possible to locate an existing outlet box on baseboard by painting the edge of the box with lipstick, and then pressing the back of the baseboard against the box. This leaves an outline of the box on the back of the baseboard. Make sure you clean up the edge of the electrical box or you'll likely get lipstick on clean wood, complicating the finish process (and irking the electrician).

**Baseboard Meets Heat: Two Register Solutions**

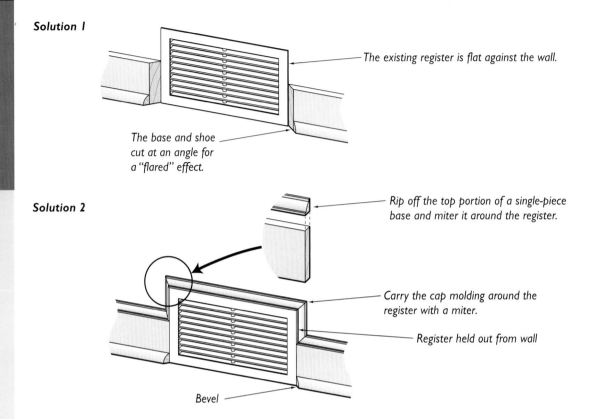

Solution 1

The existing register is flat against the wall.

The base and shoe cut at an angle for a "flared" effect.

Solution 2

Rip off the top portion of a single-piece base and miter it around the register.

Carry the cap molding around the register with a miter.

Register held out from wall

Bevel

then overlap the flush joint and return all the way to the first riser. If the stringer is thicker than the base stock, you can butt the baseboard, creating a reveal joint. In this case, the shoe molding should be cut at an open miter at the joint instead of continuing to the stair riser.

**Dealing with outlets and heat registers.** In many older houses, and in some new houses with tall baseboard, it's not uncommon for electrical outlets to be located on the baseboard. In this case, first make sure that the electrical box sticks out far enough to accommodate the extra thickness of the baseboard. Cut the base to length, and then mark the location of the box (see the sidebar on p. 65).

Wall heating registers are often taller than baseboard, creating an awkward trim detail. The easiest solution is to install the heat registers flush to the wall and then cut an open bevel on the end of the baseboard to create a flared effect. In this case, the long point of the open bevel should just touch the edge of the finish grille. The shoe molding should also have an open bevel on it, with the long point meeting the short point of the baseboard cut.

Another design solution is to install the heat registers a baseboard thickness away from the finished wall. In this case, the finish grille of the heat register laps over the square-cut ends of the baseboard and the base cap wraps around the sides and top of the register. If one-piece baseboard is used, rip the top edge off a length of baseboard and use it to wrap the register, mitering it where it meets the baseboards and at the two outside corners.

**Attaching trim to masonry.** Commonly, masonry is wrapped with wood baseboard when a brick chimney rises up through a corner of a room. In this case, I usually prebuild the corners—

securing the outside miter separately—and then attach the two pieces as a single unit. Make sure that you attach the baseboard only to the floor.

Sometimes it's not possible to prebuild a large section (for example, along the length of an exposed brick wall). In that case, attach the baseboard only to the masonry. Frequently, it's possible to nail through masonry joints with standard finish nails. If not, sink in a couple of masonry anchors, and then screw the baseboard to the wall.

Countersink the screws so you can plug the heads. You'll be surprised how few screws hold baseboard in place, especially if you use a few daubs of construction adhesive on the back of the baseboard to help secure it. Shoe molding can be attached to the floor to help hold the baseboard in place.

## Bending baseboard

Occasionally, you may have to bend baseboard around a curved wall. The easiest way to accom-

| Bending Radii for Dry Plywood or Dry Panel | | |
|---|---|---|
| **Panel Thickness** | **Bent Across Grain** | **Bent Parallel to Grain** |
| ¼ in. | 2 ft. | 5 ft. |
| ⅜ in. | 3 ft. | 8 ft. |
| ½ in. | 6 ft. | 12 ft. |
| ⅝ in. | 8 ft. | 16 ft. |
| ¾ in. | 12 ft. | 20 ft. |

Note: The radii shown are minimums; an occasional panel may develop localized fractures at these radii. (This chart was adapted from APA.)

plish this is with multiple layers of thin plywood, with each layer individually bent to fit the curve. By building up several layers, you can produce a standard baseboard dimension. Then cover the exposed top edges of the plywood with base cap.

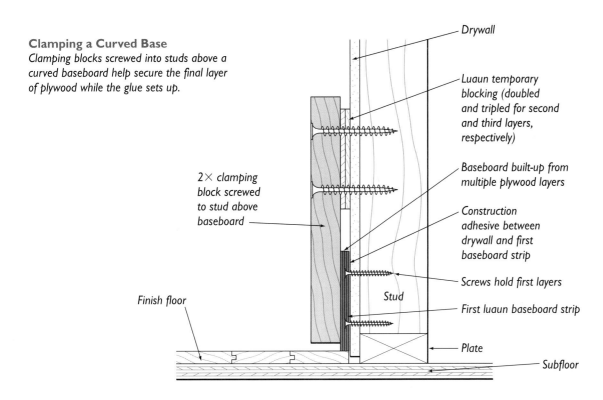

**Clamping a Curved Base**
Clamping blocks screwed into studs above a curved baseboard help secure the final layer of plywood while the glue sets up.

2× clamping block screwed to stud above baseboard

Finish floor

Drywall

Luaun temporary blocking (doubled and tripled for second and third layers, respectively)

Baseboard built-up from multiple plywood layers

Construction adhesive between drywall and first baseboard strip

Screws hold first layers

First luaun baseboard strip

Stud

Plate

Subfloor

## PRO TIP

*I always glue miter joints with yellow carpenter's glue before nailing them together. This helps keep the joint from opening up.*

## IN DETAIL

When bending baseboard to fit a curved wall, I use a hardwood-veneered luaun for the outermost lamination—even for painted trim. Standard luaun face veneers have an open, porous surface that telegraphs easily through the painted surface, creating a rough surface that must be sanded and repainted to fill the pores. Also, luaun (a relative of mahogany) has a high tannin content, which bleeds through primer and paint, creating an unsightly stain. All of these difficulties can be avoided by using a hardwood-veneered finish material.

## WHAT CAN GO WRONG

All luaun is made for exterior use, and wetting it will not affect its strength once it has dried out (though some face checking may appear). This is true of all exterior plywoods. Don't wet hardwood plywoods, however. They are made with an interior glue that isn't waterproof, so they'll delaminate.

**Base Cap for Curved Trim**
*Standard base caps can be used to cover the tops edges of the luaun layers of a curved baseboard. If the width doesn't exactly match, the cap may need to be ripped down slightly. Run the base cap through a table saw, taking care to hold the back face firmly to the fence.*

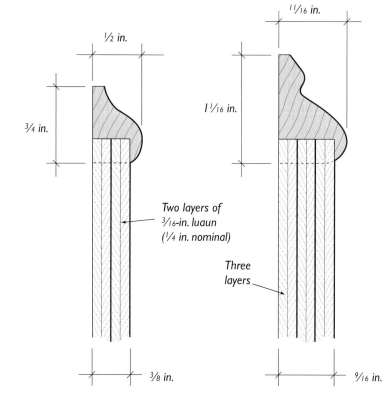

½ in.

¾ in.

¹¹/₁₆ in.

1¹/₁₆ in.

Two layers of ³/₁₆-in. luaun (¼ in. nominal)

Three layers

³/₈ in.

⁹/₁₆ in.

**When filling nail holes, take a daub of wood putty on the end of a 1-in. putty knife, fill the hole, and use the knife to scrape as much putty as possible off the surface. After the putty has dried, sand lightly with 100-grit sandpaper before finish painting.**

For a natural wood finish, use an appropriate hardwood plywood for the final layer.

I prefer to use luaun plywood, which is cheaper and often more readily available than ¼-in. AC plywood. Take care, however, to select good luaun, as voids within the core will make the strip snap with only a little bending. Both fir and luaun plywood curve to smaller radii when bent across the face grain rather than parallel to it (see the chart on p. 67). Therefore, it is often easier to use several short sections cut across the 4-ft. dimension than it is to force a long run into place. Just be sure to stagger the joints from one layer to the next.

For tight curves, you need to soak the plywood. To wet the plywood, stack wet towels on each strip. Use warm water and let the strips soak for several hours, rewetting the towels as necessary. Keep the area warm, especially when bending the piece into place. Wood fibers are elastic when warm but brittle when cold.

For a convex curve, begin at one end (for a concave curve, begin in the center) and screw the strip securely to a stud (don't set the screws too deep or they will pull through). As you wrap the strip around the curve, fasten it in a zigzag pattern to the studs and screw high and low. For the next layer, reverse the pattern—going low, then high. Paint each layer with glue before applying the next layer. Wet plywood does not affect the overall strength of water-based glue (I use aliphatic resin—a standard yellow carpenter's glue), but the setting and drying times are significantly longer. Finally, glue the finish layer, securing it in place with nails, not screws. Clamp it with a block on every stud. The nails prevent this layer from slipping around (see the drawing on p. 67).

Stock cap and shoe moldings bend quite easily. Soak the molding in water if you have some particularly tight curves. A rabbet in the bottom edge of the molding helps keep the plywood edge hidden should the molding move during seasonal changes (see the drawing on the facing page).

## Finishing up

Before painting baseboard, caulk all of the small gaps. Gaps are especially conspicuous along the base cap and shoe moldings, which twist to conform to floors and walls at a rate different than the underlying baseboard. For painted trim, run a thin bead of a latex acrylic caulk along the top of the shoe molding, under the base cap, and along any open inside or outside corners. Wipe the joints down with a damp sponge to smooth out the caulk and push it into the gaps. Also, fill the nail holes with wood putty while you are still wearing your knee pads.

## Base Meets Stringer

The finish stringer on this set of prefab stairs doesn't match the baseboard height specified for the rest of the house. Bob Smith of Housesmith Construction in Burlington, Vermont, resolved the problem this way:

**1.** Position rough stock to mark the length and where the stringer hits the end of the baseboard.

**2.** Using a tri-square, draw where the stringer line continues across the face of the baseboard up to the profiled cap.

**3.** Draw a reverse miter for the profiled cap.

**4.** Cut opposing angles and nail it home. A cap is added over the stringer to meet the reverse miter on the baseboard.

Crown

# CHAPTER FIVE
# Molding

**N**othing dresses up a room like crown molding does. Whether you install a brightly painted flat band or introduce fine shadow lines with a profiled molding, crown draws the eye upward and establishes a stylistic tone for the entire room.

Angled between the ceiling and the wall, crown molding inhabits a uniquely three-dimensional space. Although this three-dimensionality adds to the visual appeal of crown, it also makes it one of the most demanding pieces of trim to install. When the room turns a corner, the molding makes two turns—one along the ceiling and one along the wall, requiring compound cuts to join the angles. Plus, crown molding is subject to the shifting house movements acting on both planes, so it must be nailed securely to a flat, stable surface to keep the joints crisp and tight over time.

## TRADE SECRET

Most saws will cut an accurate compound angle once, and a small fraction of a degree off the angle will be hard to detect. But even slight discrepancies can be ferreted out using master craftsman David Crosby's octagon test. Cut an octagon made of equal-length segments. The sixteen 22½-degree miter cuts required to complete an octagon magnify any slight error in each angle, and the final cut will tell you all you need to know about the accuracy of the saw.

## TRADE SECRET

To speed coping, use a jigsaw to remove the bulk of the material, and then smooth the joint with a rotary sander. However, moving a jigsaw at all the crazy angles you need to cut a cope can be difficult and dangerous. The Collins Coping Foot (see Resources on p. 167) makes this task a little safer. The $29 stainless steel add-on fits all brand-name jigsaws. With DeWalt® jigsaws, an extra metal plate (provided by Collins Quality Tool®) is required.

## Simple Crown Profiles

### Band Molding

*In essence, band molding is upside-down baseboard; however, it is more noticeable because it isn't covered up by furniture. Although simple in design, it can be tough to install if the ceiling isn't flat or level. If the ceiling isn't flat, you can install the trim a bit lower down and cover the gap with a more flexible molding, such as scotia or quarter-round.*

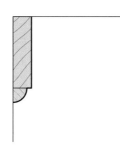

### Bed Molding

*While they are often not much more than 2 in. wide, bed moldings also have a spring angle—typically at 45, 35, or 32 degrees. They are commonly used at the top of ornate head casings to fill in around coffered ceilings or encased beams. When used in conjunction with other moldings, they are also a popular choice for large, built-up cornices.*

### Cove Molding

*Any molding with a hollowed-out curve is called cove molding. Large cove can be tough to join, because the long, sweeping inside curve can become a void if it is not cut precisely. Small cove is typically called scotia, and it is relatively more forgiving.*

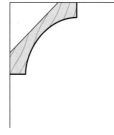

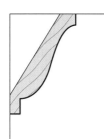

### Ogee Molding

*Standard ogee is most often what people mean when they say "crown molding."*

# Crown Types

**C**rown molding can be any angled trim that fits between the ceiling and the wall, including ogee crown, cove, and bed moldings. These profiles can be used alone as single-piece moldings or combined to create various complexities of built-up crown.

## Design considerations

In my opinion, traditional profiled crown works best in older homes with high, smooth plaster ceilings. The higher the ceiling, the larger the crown can be. If the ceiling is 10 ft. or higher, consider installing an ornate built-up cornice (see Built-Up Crown Molding on p. 87). More contemporary homes with lower ceilings, textured surfaces, or patterned tile ceilings call for simple cove or band moldings made from flat stock.

Two design principles are at work here: *contrast*—ornate moldings accent smooth surfaces and smooth moldings accent textured surfaces—and *proportion*—large moldings can overpower a small space. Many houses have a standard ceiling height of around 8 ft. (give or take a few inches) and smooth drywall or plaster surfaces. In most of these cases, standard (3-in. to 4-in.) crown works well.

## Spring angles

Crown molding always sits tilted between the wall and the ceiling. The angle of this tilt is called the *spring angle,* which is the angle at which the molding springs off the wall. Each size and style of crown molding has a different spring angle. Cove moldings often have a 45-degree spring angle, but the more common ogee-style crown moldings typically have a spring angle of 38 degrees.

Since crown molding has a complementary angle from the ceiling, it is often referred to by both angle numbers, such as 45/45-degree or 38/52-degree crown molding. Crown with a

38-degree spring angle is so common, in fact, that most manufacturers of compound-miter saws provide detents at the correct miter and bevel angles for cutting 38/52-degree crown.

## Crown materials

Like other trim materials, crown materials fall into two categories: *paint grade* and *stain grade* (see Paint Grade vs. Stain Grade on p. 23). The most common paint-grade materials include finger-jointed pine, MDF, and poplar (for custom profiles). In recent years, MDF has become increasingly more available in a variety of sizes and profiles, and it has become my first choice for paint-grade work. It's relatively stable and able to resist the expansion and contraction of temperature and humidity changes. When it is securely backed, the inside corners can be mitered and the scarf joints do not easily telegraph to the surface. This isn't necessarily the case with finger-jointed pine. Even when installed and primed well, I've been able to find finger joints after a heating season or two.

Stain-grade crown is typically run in hardwood, and because caulked joints aren't an option, it usually requires more cutting finesse. Instead of mitering the inside corners, the joints must be coped (see Cope vs. Miter on p. 74), and as you will soon discover, coping hardwood is especially challenging. You must be dead-on with your cuts, as there's no fudging a hardwood joint. In addition, a hand coping saw cuts slowly through

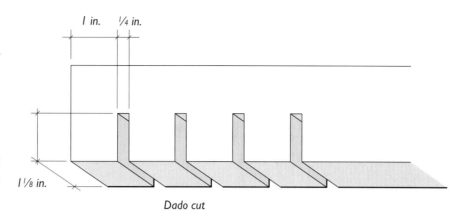

*Dado cut*

**Dentils**
*Dentils are literally the "teeth" of an elaborate cornice. The square profile looks like individual blocks set at alternating depths. However, they can be cut easily from a single piece of 5/4 lumber using a dado blade on a table saw.*

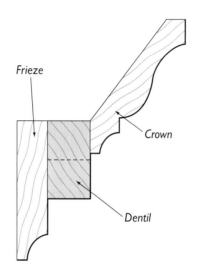

*Frieze*

*Crown*

*Dentil*

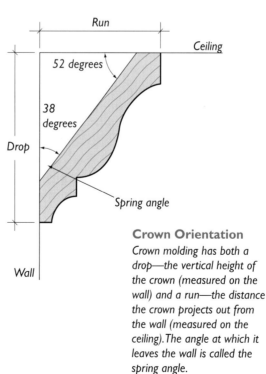

*Run*

*Ceiling*

*52 degrees*

*38 degrees*

*Drop*

*Spring angle*

*Wall*

**Crown Orientation**
*Crown molding has both a drop—the vertical height of the crown (measured on the wall) and a run—the distance the crown projects out from the wall (measured on the ceiling). The angle at which it leaves the wall is called the spring angle.*

## + SAFETY FIRST

Cordless trim saws have plenty of cutting capacity for most trim carpentry, and you can't beat them for convenience and control. But remember to play it safe: Just because these saws are light and cordless doesn't mean you can be careless with them. They are capable of hurting you just as much as a regular corded circular saw.

### TRADE SECRET

To help align molding at the correct angle in the saw, I make a simple jig with two strips of plywood screwed together at a right angle along the saw fence and table. I then apply a third strip of wood to hold the exact position for the top edge of the crown (the edge that's *down* on the saw table), as shown here.

### WHAT CAN GO WRONG

With a little caulk and paint, poorly cut joints can be made to look pretty good—temporarily. The problem is, joints filled with caulk or adhesive are likely to bulge. As the house settles or the molding expands with seasonal changes in temperature and humidity, the caulk in the joint is squeezed and then pulled apart. The result—cracking paint and lumpy joint lines—looks terrible. It's better to cut the joints accurately in the first place.

denser material; most carpenters cut hardwood cope joints with a combination of power tools. However, there's no other choice than to cope hardwood crown, which will predictably move in any heating climate, turning tight miter joints into gaping gashes.

### Cope vs. Miter

Many production carpenters would argue that MDF is stable enough to allow you to miter the inside corners and not worry about them opening up when the wood shrinks. And it's true that it is far faster to miter rather than cope an inside corner, even if the miter must be recut a couple of times to precisely fit an out-of-square or out-of-plane corner. But even if the wood doesn't shrink, coping is far from a waste of time. In fact, I cope *all* inside crown joints, whether the crown is made from expensive hardwood or MDF. Regardless of the material, a cope joint is superior to a miter joint in masking the distortions of an out-of-level and out-of-square house.

Any house undoubtedly moves. New houses certainly shrink quickly in the first few years, and then are said to "settle in." The planes that exist when you install crown molding are unlikely to be the same ones that emerge when a house settles during any given heating season. In my experience, a house never settles in permanently. Every house continually changes plane with seasonal fluctuations in temperature and humidity. Coped joints move like a knuckle, allowing these slight changes to occur without the joint opening up.

Other issues play into my decision to cope all crown joints. For one thing, I take a systematic approach to installing crown, which speeds up the job considerably. As I install the backer, I try to correct the plane, shimming out bulges and dips in both the ceiling and the wall. I also keep all cope angles in the same direction so that I

don't have to continually reset my saw. Moreover, MDF, if used, is soft and uniform, making it exceptionally easy to cope. Finish cuts can be fined-tuned with a utility knife.

# Prepping the Walls

Crown can be installed at the same time as other trim. Ideally, this is after the drywall has been taped and the ceiling has been primed but before the finish paint has been applied. In old houses, where walls and ceilings are usually not smooth or flat, it's worth spending a little time mapping the area to find the low points and out-of-square corners.

## Sagging ceilings and wavy walls

Since crown rests between the ceiling and the wall, it is subject to the imperfections in both planes. A sag in the ceiling or a wave in the wall creates a twist in the molding. Typically, ceilings are the worst. Most old houses have some kind of sag, usually a gentle curve toward the center of the room. One way to deal with it is to ignore it. Long lengths of crown are surprisingly flexible and often "wrap" around gentle bumps and shallow dips.

I always start a job by mapping the terrain with a builder's level and string to find the high and low spots. If the terrain is especially bumpy, I measure down the drop of the crown at the ends and snap a chalkline where the bottom edge of the crown should be. Then I cut the crown and tack it up, holding the edge to the chalkline. If the ceiling sag is bad, I sometimes can't fit the crown on the chalkline, and if I force it, it throws off the corner joints.

In this situation, there are three possible solutions. One is to scribe the top edge of the molding to the ceiling. To fit crown to an uneven ceiling, start by finding the low point, measure down the drop of the crown at that point, and

# COMPOUND-MITER SAWS

Compound-miter saws—commonly called chopsaws—come in two types. Both models tilt for the bevel and rotate for the miter. A *fixed-base* model has a cutting head that simply pivots from a fixed point in a manner similar to that of a standard chopsaw. A *sliding* model has a cutting head that slides on a bearing-guided bar (or two) across the table in a manner similar to that of a radial-arm saw. All sliding saws cut on the push, however, making them much safer to use than radial-arm saws. Dual compound-miter saws tilt in two directions, making it easier to cut bevels. Without this feature, you must flip the molding around, which can be quite a trick in a tight workspace.

A standard 10-in. power miter saw can only miter sanitary base and other standard 2½-in.-high base stock held vertically against the fence. And while it can square-cut wider stock (up to a 1×6), you'll need to upgrade to a 10-in. compound-miter saw or a larger 12-in. chopsaw in order to miter-cut this stock. For very tall stock, you'll need either a 12-in. compound-miter saw or a sliding compound-miter saw (which can miter-cut 5/4×10 stock or 1×12 stock). These large-capacity saws are more expensive, but they enable you to cut a wide variety of other trim stock, including shelving materials, large crown, floorboards, and jamb stock.

A compound-miter saw cuts both a miter and a bevel at the same time with the molding laid flat on the saw table. Most compound-miter saws have preset detents at the angle settings for common 38-degree crown. For crown with spring angles other than 38 degrees, you will need to calculate the angles when cutting it flat (easier said than done). Most saws include a list or chart of miter and bevel settings for a wide selection of crown. Although these saws are versatile (and expensive), you can't cut just any miter and bevel. If you have a 90-degree corner, for example, neither the miter nor the bevel is 45 degrees. Rather, the miter is 35 degrees and the bevel 30 degrees.

**Standard fixed base.**

**Sliding compound-miter saw.**

**Large-capacity saw.**

## IN DETAIL

Standard moldings can be mixed and matched to create various cornice designs. Consider the overall dimensions of the cornice. How far do you want it to drop? How far should it project into the room? After answering these questions, use any moldings available to provide the look you want. Almost any molding, including bed, picture rail, screen bead, baseboard, and flat stock, will work.

## IN DETAIL

Most crown has a slightly rounded top edge, which provides a slight shadow line along the top. If you scribe the top edge of your molding to fit an uneven ceiling, make sure that you relieve the new edge with a block plane after cutting the scribe line. This preserves the shadow line and helps conceal the fix.

snap a level line on the wall. Then cut the crown to length, tack it up along the line, and scribe the top edge. However, part of the visual appeal of most moldings is the parallel shadow lines created by the different ways light strikes the contours. So when scribing, don't bite too deeply into the top edge, or the nonparallel edge will be too easy to read. Generally, I don't scribe off more than a third of the top reveal.

Another option is to bend the crown to lie flat along the uneven wall and ceiling, but this usually means that the corners don't align as they should. In that case, the angles must be scribed. To scribe the new angle, first cut the miter and reverse bevel as you would for a regular cope. This gets you close to the angle. Tack up the crown as close as possible to the meeting piece, with the long point of the reverse bevel against the meeting piece. Holding the scribe parallel to the ceiling plane, scribe a new angle. It's difficult to get a precise profile line with the scribe, so use this line mainly as a gauge for a new miter angle, which is then cut and coped as usual.

Probably the easiest fix is to let a good drywall taper or plasterer bail you out by skim-coating (or "floating") the ceiling and walls with plaster or joint compound to conceal any gaps after the crown has been installed. Instead of scribing the crown to an uneven ceiling, you can simply install

**Rather than cut the backer to completely fill the space behind the crown, I lay it out and cut it so that the front is about ¼ in. behind the back of the crown, which allows room for positioning.**

it to a level line, leaving a visible gap between the molding and the ceiling except at the places where the crown meets the low spots. If you don't have solid backing behind the crown, you may need to slip shims into the gaps to provide a place for solid nailing along the molding edges. Cut the shims back, angling the knife behind the crown. These will later be covered along with the gaps by

**Rip backer from straight 2× materials and at a bevel to match the spring angle of the crown molding.**

Backer can be screwed to the top wall plates; it doesn't have to be pretty, just secure.

the skim coat of drywall mud, which is applied after the crown is in place.

## Points of attachment

Don't expect to march up the ladder and begin firing nails. Finding nailing in the ceilings and walls can be much more difficult than you expect. Start by figuring out how the crown molding will be attached.

At the very least, plan to spend some time mapping the stud and joist layouts. In a new house, you can fly as soon as you know the nailing, and it's usually on even centers, so it's easy to map. This isn't necessarily so in an old house where stud and joist layouts can be on uneven intervals, and weak plaster and wavy drywall create soft spots that defy secure nailing. Also, nailing into stud and joist layouts means that the crown is secured along only the top and bottom edges—the thinnest and least stable portions of the molding. Inevitably, the edge will either split or the relatively small finish nails won't find pur-

## Backer Layout

To find the dimensions for ripping 2×4 backer, draw a section of the crown on a clean piece of plywood. First, lay a framing square on the plywood to represent the wall and ceiling of the room, and then hold the end of scrap crown molding exactly as it will lie on the wall. Trace the line of the backside of the crown and the angle of the framing square. This gives you a precise layout from which to pull your numbers for ripping the backer. When complete, the backer should be undersized by about 1/4 in. from the backside of the crown.

chase and pull through the plaster or drywall. I have learned to avoid such aggravations by installing a rough backer first.

**Backer.** For standard crown, I typically make backers out of straight 2× materials ripped at a bevel to match the spring angle. Rather than cut the backer to completely fill the space behind the

## IN DETAIL

Depending on the cornice configuration, backer can be built from 1× and 2× material or from plywood for wider cornice moldings. The backer doesn't need to be continuous; it can be made of individual blocks spaced on 16-in. to 24-in. centers. Also, the backer doesn't need to wrap continuously around outside corners—two blocks on each side is sufficient. However, I prefer to build right-angled blocks for the inside corners to provide a stable base for coped joints.

## TRADE SECRET

When dealing with walls and ceilings that are way off, one approach doesn't always solve the problem. In that case, try using a combination of solutions, scribing and skim-coating as necessary to make the crown molding fit correctly.

crown, I lay it out and cut it so that the top edge is ⅛ in. to ¼ in. shy of the back of the crown. This provides a gap behind the crown, allowing me to adjust the position, if necessary. The crown still firmly contacts the ceiling and wall at the edges, but it is pinned into the meaty belly with a stout finish nail.

On at least two walls in every room, ceiling joists run parallel to the wall framing, so there may not be nailing in the ceiling. Some partition walls may have some cross blocking, which I try to pick up. But mostly I screw the backer into the studs and top plate, driving 3-in. screws at an angle into the top plate. Don't be shy about riddling the rough backer with screws to hold it in place. It doesn't have to look pretty; it just needs to be secure.

Backer isn't able to correct for all the dips in a ceiling or wall, though it does get most of them,

and it helps hold the entire assembly of pieces more securely to the structure. Still, there's no way to avoid some gaps between the molding and the ceiling or wall. For paint-grade work, you can caulk the gap if it's less than ⅛ in. For wider gaps, or gaps around stain-grade crown, float drywall to fill the gap.

# Installation Strategy

Before I actually start chopping up crown, I like to scope out the installation order, much in the same way that I do for baseboard.

### First piece same as the last

To avoid having to cut a double cope (coped joints on both ends of the same piece of molding) and to eliminate the time and fuss of resetting my saw for each cut, I set up all my cuts in the same direction. Usually, I start with the trim on the

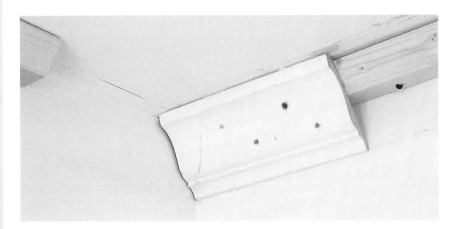

A starter block is temporarily screwed to the backer.

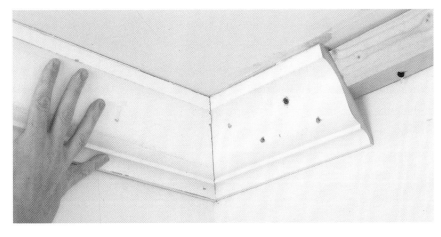

**Nail the first piece at the butt end only.**

longest wall opposite the door, making one end coped and the other end butted square into the corner. Because I'm left-handed, I typically butt right, cope left. Find what suits you and stick to it. The direction is less important than consistency so that you don't have to constantly reset the angle on your saw.

## Starter block

If you set up all your cuts in the same direction, the first piece of molding must be installed with a temporary block to fit against the coped end. This block consists of a scrap piece of 1-ft.-long crown, which is cut square to fit in the corner and temporarily screwed to the backer. I typically install this starter block, then measure and cut the first length, install that, and then work my way in one direction around the room. Nail the first piece of crown near the center toward the butt end only. This provides enough support to hold up the entire length of crown but leaves the coped end free so that you can remove the starter block and slip the last, full-length piece of crown in place.

## Measuring crown lengths

Because of the mud buildup in corners, it's often hard to get an exact reading by bending a tape measure into the corner. Even the most practiced carpenter can misjudge exactly where that bottom edge of the cope joint will land, especially when taking repeated measurements. Therefore, for most crown measurements I use pinch sticks, butting one end to the previously installed length and the other end to the wall. Pinch sticks are critical in tight, enclosed spaces, such as alcoves. Short lengths of molding aren't very flexible, so the length must be perfect.

On very long runs, it's not as critical to make a set of pinch sticks (I certainly don't take the time to make a set to measure just one long wall).

Instead, I rely on a tape measure, aiming for a measurement that is $1/16$ in. to $1/8$ in. long. You can then bend this piece slightly to get it in place. If it's slightly long, it will snap into place and the points of the back-cut will dig into the wall.

## Evaluate outside corners

If the room has an outside corner, it might make sense to start at that point. If you have more than one outside corner, consider cutting them all at once (for example, cut all of the right-hand miters and bevels at once, and then cut all of the left-hand miters and bevels). I still try to map out the cuts so that all of the compound cuts for the copes are angled in the same direction. If the legs of the outside corner are short (for example, at the top of a small bump out in a room), you may need to temporarily tack up the coped leg of the outside corner. This lets you later remove it to fit the last piece of crown that laps behind it.

Rooms with bump outs often have a few very small pieces of crown. In this case, continue to cut all the copes in the same direction, but cut the compound angle and cope the piece before you cut it to length.

# Cutting Crown

Crown molding can be especially challenging to cut because it already has one angle—the spring angle—built into it. However, if you always keep track of which edge is the bottom and which edge is the top and constantly picture how the molding will sit when it is nailed in place, you can usually puzzle through all the cuts.

As with all trim cuts, I try to group similar cuts together. That is, I first cut all the scarf joints, then all the outside miter joints, then all the inside corners. This strategy saves time with any trim. With crown molding, it's especially helpful because it helps you keep track of the three-dimensional orientation of the molding.

## PRO **TIP**

*When installing straight backer material, leave an adjustment gap into which you can push the edges of the molding. You'll then conform to the ceilings and walls.*

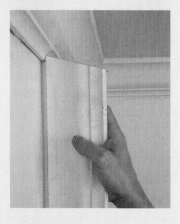

## TRADE SECRET

One way to deal with gaps between crown molding and walls or ceilings is to skim-coat, or "float," the walls with plaster or drywall mud. Carpenter David Frane taught me how to make a notched screed from a scrap of crown molding. The notch keys with the edge of the crown molding. The other end skims the surface of the wall or ceiling. I make screed about 18 in. long, which creates a long, thin taper of mud that feathers seamlessly into the existing surface.

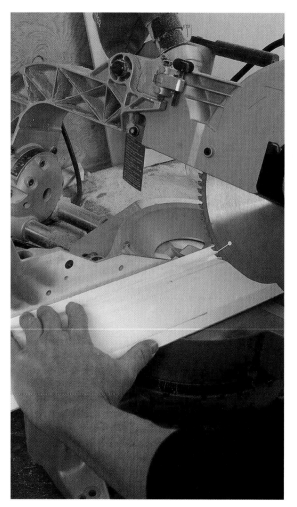

Every scarf joint begins with a compound cut on the end of one board to mate with a corresponding compound cut on the end of another.

## Scarf joints

Crown molding usually comes in 12-ft. lengths and rarely longer than 16-ft. lengths. If a wall is longer than the crown stock available, you'll need to splice, or scarf, two lengths together. I make all these scarf joints first (there usually aren't too many). In effect, I focus on manufacturing the longest lengths of molding first, and then look at the other types of cuts, tackling one type at a time. For scarf joints, I depend on two critical devices: a biscuit and a splint.

**Biscuit.** A biscuit holds the two pieces of scarf joints in alignment. To make a biscuit, cut a 20-degree-angle bevel and a 20-degree-angle miter, or a 20-20 *compound cut,* on one end of the molding, and then cut a corresponding 20-20 compound cut on the mating piece. Lay the molding and the biscuit jointer flat on the same surface (usually on a workbench or the floor) and make biscuit slots in each piece. Elevate thinner moldings slightly by placing them on a piece of ⅛-in. luaun. The slot should fall somewhere in the center thickness, not in the contoured face of the molding. The long point of the bevel holds the biscuit jointer away from the center area

A biscuit helps align the mating pieces of crown and stabilizes a scarf joint. **Because of the bevel at the end, the biscuit joiner must be set at full depth to fit a smaller #10 biscuit.**

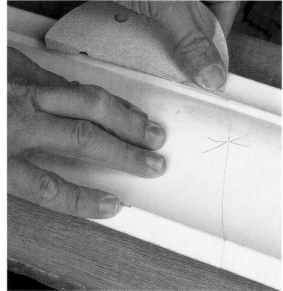

A thin plywood splint—glued and stapled across the back of the crown molding—secures the joint, creating one long piece of crown.

The completed scarf may need to be touched up with sandpaper to smooth the meeting edges.

where the slot is cut. By setting the biscuit jointer to the maximum depth, you gain enough depth to fit a smaller #10 biscuit into the slot.

**Splint.** A biscuit guarantees alignment, but it doesn't hold the end grain together well. To fasten the two pieces more securely, cut a strip of ¼-in.-thick luaun to act as a splint across the back and bridge the joints. I fasten this splint with plenty of glue and ½-in. staples in a standard T-50 staple gun. The staples hold the strip in place until the glue cures. Keep in mind that if you use backer, you must cut the splint narrow enough and hold it near the bottom edge of the molding, so that it won't get in the way, or leave out the backer where the splint will fall. After securing the splint to the scarf joint, you may need to smooth out the joint with sandpaper.

## Cutting compound miters

The method I use to cut inside and outside corners depends on the type of saw. I prefer using a conventional miter saw and a simple jig that holds the crown at an angle, similar to the way in which it will be installed. When you set the saw for a

## Interior Wood Glues

Titebond® (left) is simply a yellow glue colored brown so it won't show up on dark wood. However, I've never thought this to be a useful feature of glue, since all glue should be thoroughly wiped or sanded off before finishing and joints should be airtight, rendering glue lines invisible. For most interior work, I use regular yellow carpenter's glue, also known as "aliphatic resin" on the label (center).

Yellow glue is stronger than ordinary white glue, and it's more heat resistant, so it doesn't melt as easily under the assault of a sander and instantly gum up the paper. When I'm working in high-humidity areas, such as a kitchen or bathroom, I typically use a Type II, or cross-linked PVA, glue (right). Although not exactly waterproof (meaning that it can't be used outdoors), it's a good choice when I'm using only one type of glue. While it cleans up with water, you'll need to work quickly with it, as it is noticeably tackier and has a shorter drying time than ordinary yellow glue. As a rule, if a wood glue can be cleaned up with water, it's for interior use; if solvents are needed, save it for exterior use.

## PRO TIP

*Even after a scarf joint has been biscuited, splinted, and glued, it's still a weak spot. Support the joint until the piece is nailed firmly in place.*

## TRADE SECRET

To precisely measure the closed distance between two walls, make a crown stick with two narrow boards that slide by each other. By holding the two sticks together and sliding them until one end of each stick touches a wall, you obtain an accurate representation of the distance, which can easily be transferred to molding stock. I cut the ends of the pinch sticks to a point so that there is one positive contact point meeting the wall. Once you know the distance, mark the sticks with two lines for positive realignment. Use a spring clamp to hold the sticks together when transferring the length to the molding.

If the wall corner is slightly out of square, an outside miter may need to be back-planed to fit tightly.

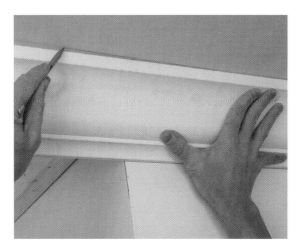

simple miter, the orientation of the molding automatically provides the bevel angle. However, this method is limited by the size of the saw. Large crown doesn't fit, unless you have a very large miter saw.

The alternative is to cut the crown flat with a sliding compound-miter saw. In general, sliding compound-miter saws are more versatile machines, but you must find the correct saw setting, which isn't necessarily obvious.

## Outside corners

Crown outside corners must be joined by compound miters. You also need good backer for outside corners. Any crown over 4 in. wide should have continuous backer behind the corner (the

Draw a line where the molding meets the ceiling near the outside corner. Then position the opposing length of molding and mark the ceiling. The intersection of the two lines represents the long point of the compound miter.

## CUTTING CROWN FLAT

When cutting crown with a sliding compound-miter saw, the molding must lie flat on the saw table. To set the miter and bevel angles on the saw, use the settings in the chart, which are based on the spring angle of the molding. Before making the cut, keep track of the top edge of the molding. Depending on the type of cut you are making, either the top edge or the bottom edge lies against the fence. The photo sequences shown here demonstrate the correct orientation for each type of cut. (*Note:* With a dual compound-miter saw, you do not have to flip the molding around like this. Instead, simply tilt the cutting head in the opposite direction.)

### Coped Inside Corner

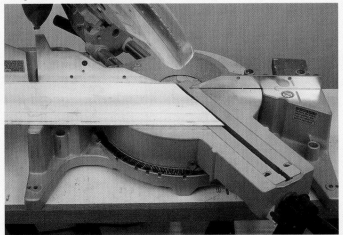

**Left-hand piece: Top edge against fence.**

**Right-hand piece: Bottom edge against fence.**

### Mitered Outside Corner

**Left-hand piece: Bottom edge against fence.**

**Right-hand piece: Top edge against fence.**

## Angle Settings for Sliding Compound-Miter Saw

| Type of Crown (spring angle in degrees) | Miter (angle on table in degrees) | Bevel (tilt of blade in degrees) |
|---|---|---|
| 30 | 27 | 38 |
| 35 | 30.5 | 35 |
| 38 | 31.5 | 34 |
| 40 | 33 | 33 |
| 45 | 35 | 30 |
| 52 | 38 | 26 |

## IN DETAIL

If your house has out-of-square corners, you may want to invest in a Bosch® DWM40L Miter Finder (see Resources on p. 167). It measures any angle to within 1/10 of a degree and provides the miter and bevel angles for any crown molding. There is no way to program spring angles into memory, so unfortunately you need to repeat the simple steps for calculating bevel and miter angles for every corner. Still, this tool can save loads of time on any crown job.

## TRADE SECRET

I've noticed that shifting crown seems to push out the putty used to fill nail holes, so I started using epoxy wood putty. This material is similar to auto body putty (commonly known as Bondo®) and it holds much better than other fillers do.

backer should also be mitered). For crown less than 4 in. wide, you can end the backer at the corner and simply rely on small finish nails to hold the miter closed, provided that it was cut accurately in the first place.

If an outside corner is square (use a framing square to check it) and you have a crisp corner (drywall corner bead, for example), you can just measure along the bottom edge and cut the molding at a 45-45 compound miter. To measure the length for an outside miter, hold the stock in place and mark where the bottom edge of the molding passes the outside corner. Remember that drywall corner bead pushes the corner out slightly, so you may have to back-plane the miter to get a tight fit.

Plaster corners may be slightly rounded, making it difficult to measure the exact length along the crown's bottom edge. If the walls are reasonably square to one another (and only the very corner is rounded), mark the ceiling where the top edges of the molding meet. This lets you measure the length along the top edge to the long point of each miter cut. Position a length of crown along one side of the outside corner and draw a line where the molding meets the ceiling near the outside corner. Then position the opposing length of crown along the other side of the corner and mark the ceiling. The intersection of the two ceiling lines represents the long point of a 45-45 compound miter on each length. After cutting the miters, check the fit of the outside corner before nailing it in place.

In an old house, the corner may be out of square, so the miter will not be a 45-45 compound cut. This situation requires a more elaborate method to find the exact miter angle (see the sidebar on the facing page).

## Inside corners

You can cut inside corners in two ways: with a cope or with a miter. As mentioned earlier, I prefer to cope all crown molding. However, MDF materials can be joined with a compound miter, which is the fastest and simplest cut. Cut both pieces to equal length with mirror-image

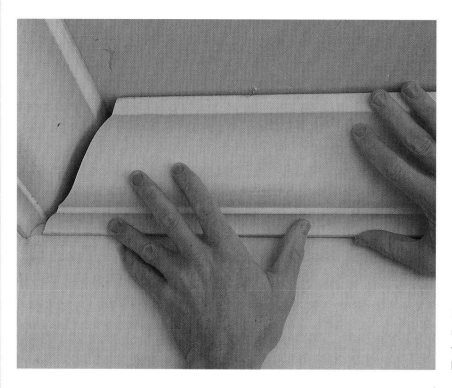

**Finishing a cope may require several trial fits to make sure that it is back-cut far enough.**

miter/bevel cuts. I think of mitered pieces as equal mates—both meet in the middle.

A coped joint, on the other hand, is a marriage of complete opposites. One length of molding is cut square and butts tight to the wall. The second piece—the coped end—"keys" to the profile of the butted piece at a right angle.

**Inside miter.** To cut an inside miter, measure the length of the molding along the bottom edge. Remember that the long point of all inside miters will be on the bottom edge. Set up the molding exactly as you would for an outside miter but cut the opposite, or *reverse,* bevel.

For inside miters to fit well, the wall and ceiling planes must be nearly square and reasonably flat. Otherwise, the molding will twist and the inside miter won't fit tightly without a lot of fuss.

**Coped joint.** A coped inside miter starts with a compound miter—the same one used for an inside miter (see the top left photo on p. 86). The contoured edge along the face of the cut defines the profile to be coped. If you're cutting with a standard chopsaw, set the molding in the saw just as you would for an outside miter. However, rotate the table so that you cut the opposite, or reverse, bevel. This exposes the face of the crown, giving you the exact shape of the profile to cut with the coping saw.

Next, mark the edge of the profile with the flat of a pencil lead, highlighting the contoured edge so that you can cut along the line with a coping

## Bisecting an Angle

Carpenter, boat builder, and writer Jim Tolpin taught me a simple method of measuring the exact bevel setting for an out-of-square corner using a *bevel board*. The bevel board is nothing more than a piece of wood with some very carefully drawn angled lines ranging from 0 to 90 degrees. I made mine using a simple protractor and a very sharp pencil. To get the exact bevel angle for an out-of-square corner, start with two short scraps of wood that overlap at the corner.

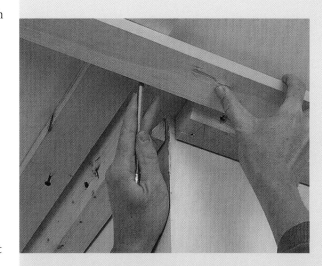

1. Mark the long point of the angle on the underlapping piece (see the top photo).

2. Mark the short point on the same piece (see the center photo).

3. Find the angle on the marked piece with a bevel square. Then lay the bevel square on the bevel board and read the angle (see the bottom photo).

## PRO TIP

*Large rooms with high ceilings call for much larger moldings; at this size, they are commonly referred to as "cornice moldings."*

## TRADE SECRET

I use a standard 10-in. miter saw for cutting crown that is less than 4 in. wide. In fact, I prefer this saw because I only need to worry about the miter angle. By setting the molding on the saw at the spring angle, a single miter automatically cuts the bevel. The only drawback is that the saw has a limited depth of cut: Large molding doesn't fit in the saw at an angle. For that you'll need a larger miter saw or cut molding flat with a compound-miter saw.

## IN DETAIL

To identify a good coping saw, look for one with a stiff back. The back holds the blade in tension, and the more tension, the easier it is to steer the blade through the twists and turns of coping. The best ones I have seen come from Garrett-Wade (see Resources on p. 167). But in truth, I typically use an inexpensive Disston® that I picked up at the lumberyard. I have gotten good results with it, though occasionally I have to gently spread the bow a bit to maintain good tension on the blade.

Every cope for an inside corner of crown molding starts with a miter and a reverse bevel to expose the profile.

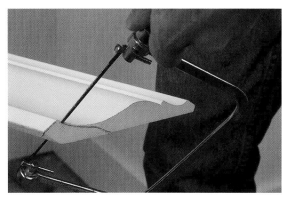

After cutting a reverse bevel, cut the profile with a coping saw.

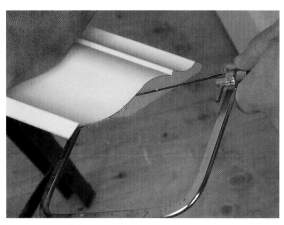

Back-cut by 90 degrees.

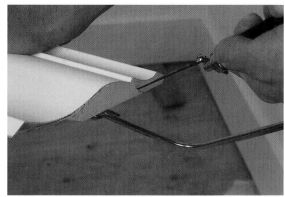

Saw from different directions to keep from breaking the edge.

Smooth with a rasp.

Final edge.

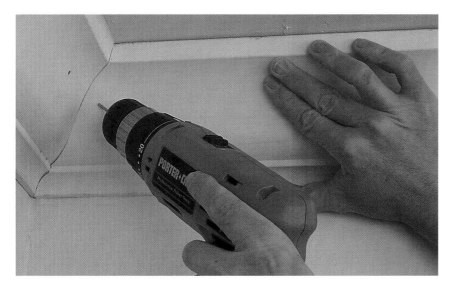

**When the cope fits tightly, predrill holes in the meaty portion of the crown's profile before nailing it.**

saw. I always back-cut slightly by tilting the coping saw past 90 degrees so that the meeting piece of molding can "pass by" cleanly. The coped piece should intersect only along the profile line. When cutting with a coping saw, I typically saw toward the curve from different directions so that I don't have to twist the saw and risk breaking the fragile edge of the profile. After cutting out the bulk of the material, smooth out the back-cut with a rasp, until you have a smooth, sharp feathered edge. To install the coped piece, predrill holes to avoid splitting the piece.

# Built-Up Crown Molding

So far, all of the techniques I have described in this chapter apply to single-piece crown moldings, ideally from 2½-in. to 4½-in. wide. Large rooms with high ceilings call for much larger moldings; at this size, they are commonly referred to as "cornice moldings." While it's possible to have large one-piece crown custom milled, a much more stable (and often less expensive) alternative is to build up large crown from several smaller profiles.

Built-up crown need not be complex—it can be as simple as a one-piece base and a stock ogee crown. A more complex variation on the same

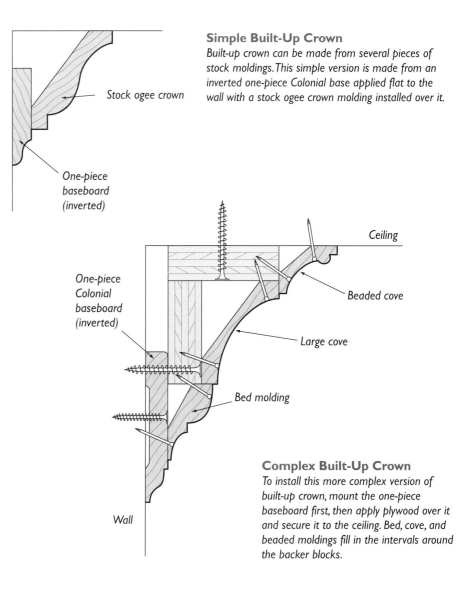

Stock ogee crown

One-piece baseboard (inverted)

**Simple Built-Up Crown**
*Built-up crown can be made from several pieces of stock moldings. This simple version is made from an inverted one-piece Colonial base applied flat to the wall with a stock ogee crown molding installed over it.*

Ceiling

One-piece Colonial baseboard (inverted)

Beaded cove

Large cove

Bed molding

Wall

**Complex Built-Up Crown**
*To install this more complex version of built-up crown, mount the one-piece baseboard first, then apply plywood over it and secure it to the ceiling. Bed, cove, and beaded moldings fill in the intervals around the backer blocks.*

## PRO TIP

*When wood is removed from one side of the board, large crown or cove can cup and twist. A large cornice built from smaller components is more stable.*

## IN DETAIL

It's not uncommon to mangle or even snap a blade when coping, so keep a healthy supply of replacement blades on hand. Each blade has a single row of jagged teeth oriented in one direction, and most saws allow you to mount the blade facing both directions—with the teeth either angling toward the handle or away from it. Hence, carpenters often debate whether one should cut on the upstroke or the downstroke. I don't think it matters—use whatever feels right.

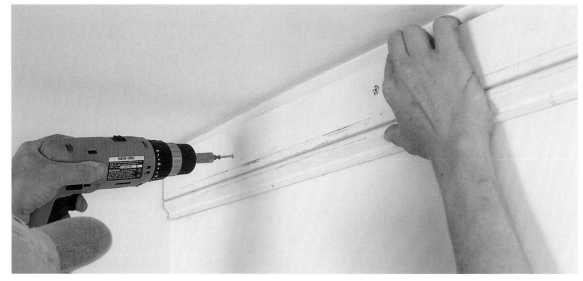

Snap a line on the wall to represent the overall drop of the cornice, and then install a piece of inverted baseboard on that line.

idea also uses standard base stock and standard crown moldings, though with the addition of backer to help hold the various pieces of molding in the right position. There are a range of crown variations that are possible utilizing the important principles of installing backer and locking moldings together.

## Cornice backer

The use of several small profiled pieces calls for a stable nailing base. Backer isn't just an option—it's a requirement with a large cornice. Depending on the complexity of the cornice, the backer can be ripped from plywood, fashioned from stock baseboard molding, or made from combination of the two. The wide, flat area of baseboard molding can provide the backer for other moldings, and the profiled cap on the molding can be left exposed as part of the finished cornice.

Once I've come up with a cornice design that I like, I lay out the pieces and decide where the different moldings will spring off the finish backer. First, I snap a line on the wall that represents the overall drop of the cornice, and then install a piece of standard Colonial baseboard molding stock on that line. I install the base stock upside down, with

the profiled cap hanging down, and screw through the middle of the wide, flat area of the molding with screws that are long enough to penetrate at least 1 in. into the studs or top wall plates. I also draw a line on the finish backer to represent the bottom edge of the next molding to be applied. I try to do this before installing the backer so that I can use a tri-square to scribe the line parallel to the finish edge.

If I'm using a finish backer along the ceiling, I snap a line on the ceiling to represent the overall run of the cornice, and then install the profiled top of the baseboard stock to that line. When running the second piece of backer, I keep a piece of the crown with me so I can periodically check that the two backer pieces are parallel and that the reveals are consistent. Otherwise, I install the crown to the line drawn on the first piece of backer, first checking the fit with a piece of scrap crown. If the cornice requires plywood backer blocks (which I make up in advance from scrap plywood), then I install them at each joist, if possible.

I try to drive at least one screw through each flange: one into a ceiling joist and one into a wall stud. However, on walls that run parallel to the

Check the fit of the top ceiling board with scrap crown molding.

Install the crown molding, starting with a long section butted into a corner.

Backer blocks made from scrap plywood provide a secure nailing base for large crown molding.

After the backer is in place, you are ready to install the molding, starting with the first layer of bed molding.

ceiling joist where I don't have any nailing in the ceiling, I drive several smaller screws through the wall into the finish backer, relying on the rigidity of the plywood backer blocks to hold up the cornice. After the backer is in place, the subsequent moldings can be installed one layer at a time, starting from the bottom.

Don't glue the individual moldings together. They should be free to move independently of each other to accommodate changes in temperature and humidity. However, you may want to use a bead of adhesive latex caulk, such as Phenoseal, to hold the top molding to the ceiling.

# Wainscoting

# CHAPTER SIX

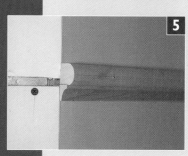

Traditionally, the term "wainscoting" has referred to any kind of wood wall paneling, though today it usually means paneling over the lower portion of a wall. Wainscoting was a common interior feature in American homes from the mid-1800s into the twentieth century. Chair rail, which caps most wainscoting, once served a practical as well as a decorative purpose, protecting the wall from chairs set up against it. For Romantic- and Victorian-era homes, wainscoting is a period detail, but it adds traditional warmth and elegance to a contemporary house design, too.

We'll take a step-by-step look at the simplest wainscoting method, which uses vertical tongue-and-groove boards. Vertical board paneling is simple, economical, and quick to install, so it is the type that I install most often. Then I'll introduce a modified frame-and-panel system. Generally, such systems require more joinery, but the method shown here is much simpler, and it gives a good approximation of true raised panels.

## PRO TIP

*Wainscoting typically covers the lower third of a wall. Start with the one-third rule, but match the height to the existing trim or other interior details.*

### Creating a Golden Rectangle

*Step 1: Draw a square the desired width of the rectangle.
Step 2: Bisect the baseline (point A) and extend a compass from A to the opposite corner (point B).
Step 3: Swing an arc from B to the baseline of the rectangle (point C), which is the length of the Golden Rectangle.*

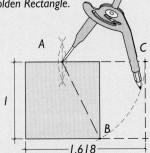

### IN DETAIL

The *golden mean*—a rectangular proportion of five to eight—is a useful design rule that is used often in construction. When I'm planning frame-and-panel wainscoting and windows and casing aren't defining factors, I use the golden mean to establish the heights and widths of the grid. To preserve this proportion, the panels of tall wainscoting often run vertically.

# Designing Wainscoting

**Because it covers** such a substantial portion of the wall, wainscoting taxes the limits of wood's dimensional stability. Over a large area, the cumulative movement of many boards joined together can lead to substantial problems with cracked boards and open joints. Thus, any wainscoting design must account for the inevitable movement of wood that occurs with changes in humidity and temperature.

The simplest traditional method uses many relatively narrow tongue-and-groove boards. Each board is nailed along the tongue edge so that the groove edge is free to move. As the boards expand and contract, the groove slips over the tongue, and the overall movement of wood across the entire wall goes undetected. The other traditional

**Bead-board wainscoting is the simplest to install and just as elegant as raised-panel wainscoting.**

approach uses molded, solid-wood panels that essentially float in a frame fixed to the wall.

## Tongue-and-groove wainscoting

Also known as vertical-board wainscoting, tongue-and-groove wainscoting looks best in simpler house styles. It works well with country and Colonial design schemes. With a simple, square-band chair rail, it can also fit in contemporary schemes designed to accentuate clean vertical and horizontal lines.

Tongue-and-groove wainscoting comes as individual boards sold in random widths, in standard widths from 2½ in. to 5 in., and in thicknesses ranging from ⅜ in. to ¾ in. Most home centers sell shrink-wrapped bundles of the thinner ⅜-in.-thick material in precut lengths from 3 ft. to 4 ft. Lumberyards often sell wainscoting in various thicknesses by the linear foot.

Although just about any tongue-and-groove boarding can be used for wainscoting, material sold expressly for this purpose typically has a milled profile—either a simple beveled edge, called V-groove, or a beaded edge, called bead board. Wider widths often have an extra V or bead milled into the face as well. Some traditional-minded carpenters consider only bead board to be true wainscoting material and use V-groove as a finish ceiling board (common for porch ceilings). Combination material with a V-groove on one side and bead on the other is also available, though I have only seen it in ¾-in. thickness.

## Frame-and-panel wainscoting

Frame-and-panel wainscoting seems most at home in the more elaborate interiors of Victorian and Queen Anne-style houses. True raised panels deal with the problems of cumulative wood movement as well as any woodworking method can. Yet, as ingenious and effective as this approach is, it is exceptionally difficult and time-consuming

## Chair Rail Profiles

### Square band

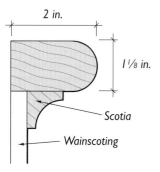

1 1/8 in.

1 1/2 in.

Wainscoting

### Nosing

2 in.

1 1/8 in.

Scotia

Wainscoting

### Bead band

2 in.

3/4 in.

Bead

Door stop

## Wainscoting Board Profiles

### Bead board

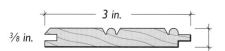

3 in.

3/8 in.

### V-groove

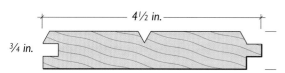

4 1/2 in.

3/4 in.

**Modified frame-and-panel wainscoting with recessed panels.**

**True raised-panel wainscoting.**

**PRO TIP**

*The base of wainscoting looks best when it matches the height of other baseboards. Try to echo other baseboard heights when they can be seen prominently.*

**IN DETAIL**

Drywall installers often use a small router to speed the task of cutting holes for electrical boxes, plumbing lines, and heater ducts. This tool is indispensable for cutting plaster or drywall to install backer for wainscoting. I use a Roto Zip tool (see Resources on p. 167), which uses special bits to zip through both drywall and plaster and lath. Have a helper follow the cut with a vacuum nozzle to eliminate dust.

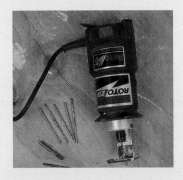

**IN DETAIL**

The traditional way to provide nailing for wainscoting is to install 2× blocking between the framing. However, I prefer to install ½-in. or ⅝-in. plywood in place of drywall on the lower portion of the wall. This provides a more solid, continuous nail base.

to execute well. I put raised panels right up there next to dovetails as one of the most challenging woodworking maneuvers.

It's possible to reproduce the look of frame-and-panel wainscoting with modern materials and simplified joinery. Sheet goods, such as plywood and MDF, are stable enough to allow you to reproduce traditional details using simple methods. The result may not be true frame-and-panel construction, but it will look elegant nevertheless. And more important, with a minimal investment in time and expertise, the paneling will remain beautiful for as long as the house stands.

## Planning your installation

Wainscoting varies in height, but as a general rule, it covers the lower third of a wall. So, for a typical 8-ft.-high ceiling, aim for wainscoting between 30 and 36 in. tall. I try to align the top rail with other features in the room. Countertops are typically 36 in. tall, bars are often 42 in. tall, and desks and tables are typically 30 in. tall. All of these features make acceptable wainscoting heights (including chair rail). In rooms with tall

ceilings, wainscoting even looks good at 60 in. or taller.

Similarly, the base of wainscoting looks best when it matches the height of other baseboard in the house. I generally try to echo other baseboard heights when they can be seen prominently through doorways or around corners. But, like all design rules, this one can always be broken, as long as you do so deliberately.

The top rail, or chair rail, is often the most prominent feature of wainscoting. Choose a profile for the top rail that matches, or in some way conforms to, the existing trim. If window and door casing is flat and plain, I aim for a simple square-edged band. More elaborate profiled trim calls for a built-up rail fashioned from several pieces of profiled molding.

With a frame-and panel system, the rectangle becomes the primary design element, so you want to pay attention to the alignment, proportion, and orientation of these defining rectangles. I typically match the rail width to the casing width, so that the vertical band of the casing visually matches the stiles on the wainscoting. In general, panels

**Tall wainscoting, applied to just one wall, requires minimal work but provides a distinctive interior touch.**

**The stiles of this frame-and-panel wainscoting mirror the width of the door casing.**

should be proportioned to match, or echo, other architectural elements, such as window sash, door panels, or built-in cabinets. However, this proportional unit provides only a starting point, as it's often necessary to alter the width of panels in order to evenly divide wall areas.

As with all interior trim, a great deal of the design focuses on the details, such as the way the chair rail intersects with the door and window trim. These details can have a considerable impact on how long the job will take, depending on how simple or elaborate they are.

# Installing Backer

All wainscoting needs solid backer in the walls for nailing. In new construction, many carpenters install blocking between the studs—usually three rows of 2×4 blocking installed "on the flat" near the wainscoting's top, bottom, and center. However, even in new construction, I prefer to install plywood backer over the studs, which pro-

**The panel sizes of wainscoting should be planned carefully so they evenly divide wall areas.**

## PRO TIP

*When working into a corner, keep a Surform® tool handy to level off extra drywall compound.*

## WHAT CAN GO WRONG

Don't glue wainscoting to drywall. Sometimes, bundled wainscoting is accompanied by "instructions" that describe how to install the boards directly to drywall using construction adhesive. Voilà, instant wainscoting. Wrong! Gluing solid wood to an immovable surface is never a good idea. Tongue-and-groove material is designed to move. Nail only one edge of the board so that the other is free to move—the floating groove slips over the stationary tongue—as the boards expand and contract with changes in temperature and humidity. Chances are high that boards glued to plaster or drywall will split during the first heating season.

## WHAT CAN GO WRONG

It's not unusual to have a buildup of drywall compound at a messy joint or loose or sagging plaster on an old wall. When wainscoting is installed over a wall with little clearance, the boards will bulge out of alignment unless you shave excess drywall mud or chip back the plaster—a messy and aggravating job to say the least.

**Modified frame-and-panel wainscoting begins with continuous plywood backer in place of drywall on the lower portion of the wall.**

vides continuous nailing for individual boards and reduces air leakage.

The many cracks between individual boards in both vertical-board and frame-and-panel wainscoting create potential places from which air can leak to exterior walls. Unless you make an extra effort to keep the walls tight, the many air leaks can allow warm air to escape, and in cold climates, create moisture problems inside the walls. The best way to avoid this problem is to replace the drywall behind the wainscoting with plywood, and then seal the top gap between the plywood and the drywall with expanding foam. If you opt to strip out the drywall to install horizontal nailers, seal the gaps with foam above and below each nailer on all exterior walls.

Retrofitting a nailing base, or backer, in existing construction requires considerably more work. The most economical alternative is to cut out the drywall in strips and let in horizontal nailers (see the photos on p. 98). A better alternative—and one that provides continuous and more

stable nailing—is to remove the drywall or plaster and lath entirely behind the wainscoting and replace it with plywood.

## Backer thickness

The thickness of the backer depends on the thickness of the original wall surface. When installing continuous plywood backer, try to match the thickness of the wall covering (drywall or plaster), so that the wainscoting is applied to the same plane and projects forward. For let-in strip backer, I aim for backer that's slightly thicker than the wall, allowing the applied wainscoting to bridge any imperfections in the surface. This means that there will be a small gap between the back side of the wainscoting and the old wall, which will be covered at the top by the chair rail. For example, I use 5/4 material for let-in backer in walls covered by plaster and lath. For walls covered with ⅝-in. drywall, I use ¾-in.-thick (1×) nailers, and for ½-in. drywall, I use strips of ⅝-in.-thick plywood. All of these provide a nailing surface that sticks out slightly from the original wall.

When determining the thickness of the backer, check the thickness of existing baseboard and window and door casings. There may not be room to provide ample clearance for wainscoting and also preserve a nice reveal along the baseboard or window and door casings. In this case, you want to have at least a ⅛-in. reveal showing along the edge of the trim. If you're installing new trim, consider using 5/4 casing and perhaps shimming out the baseboard, so that these pieces of trim project well out from the surface of the wainscoting.

## Backer layout

To install let-in nailing strips or continuous sheet backer, start by establishing a level line that represents the top of the wainscoting. I snap a chalkline around the room, usually with blue chalk, which is water-soluble and can be washed off if the walls

**Wainscoting Backer**

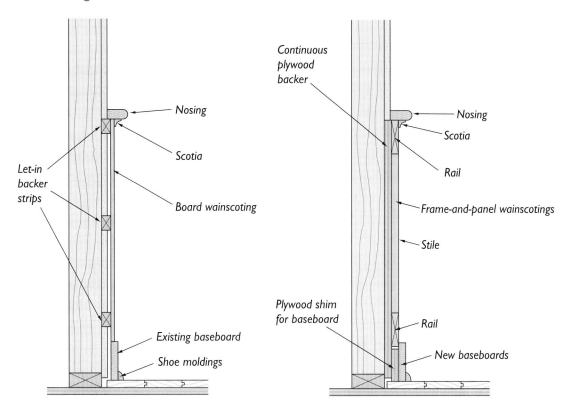

Nosing

Scotia

Let-in
backer
strips

Board wainscoting

Existing baseboard

Shoe moldings

Continuous
plywood
backer

Nosing

Scotia

Rail

Frame-and-panel wainscotings

Stile

Plywood shim
for baseboard

Rail

New baseboards

**Plywood fills in below the wainscoting to shim the baseboard outward.**

won't be repainted. This line represents the top edge of the chair rail that caps the wainscoting.

Below this top line, I lay out the locations for the nailing strips. The upper nailer should be just below the top line to provide nailing for the chair rail and the top end of the wainscoting. All of these lines can be quickly established around the room with a layout stick.

When I am working off an existing baseboard, I position the lower nailing strip a couple of inches above the baseboard. The nailer needs to support only the lower end of the wainscoting. However, if I am applying a new baseboard over the wainscoting, I want the lower nailing strip to provide backer for it as well.

## Removing the old wall

Removing the old wall covering is often the most laborious part of retrofitting wainscoting. I've found that this task can be greatly simplified by

## IN DETAIL

To transfer an elevation around the room, I often use a story pole. This is simply a narrow board that helps mark the height of chair rail and backer strips. Using a story pole saves time, since I only have to pull out my tape measure once to mark the stick. Then I hold the bottom end of the stick on the floor (or the top edge of an existing baseboard) to transfer consistent heights. I use a story pole to mark both sides of each corner, window, and door, and then I snap a chalkline to establish layout lines.

**Snap a series of layout lines around the room to mark the location for let-in backer strips.**

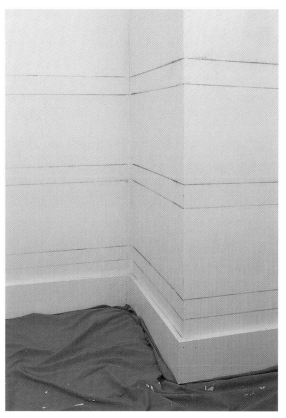

**The completed layout for let-in backer is ready for cutting.**

using a drywall cutout tool, such as Roto Zip®. Without this tool, you must cut along the layout lines with a utility knife, scoring the lines several times in order to cut completely through the old surface.

Plaster is sometimes a little easier to remove than drywall. A couple of deep score lines made with a sharp utility knife are usually sufficient to cut through the surface. Then, chip out the plaster with the claw end of a hammer and remove the wood lath strips. Be sure to pull out all the old nails from the edge of the studs so that you have a clean face on which to attach the horizontal nailers.

## Retrofitting Nailing Strips

If you install wainscoting in an old house, it's unlikely that there will be sufficient nailing. In that case, you must strip out the old plaster or drywall and let in horizontal nailers. To do this, follow the steps below:

1. Establish a level line for the top of the wainscoting.

2. Snap layout lines for the top rail and at least three nailing strips.

3. Cut through the drywall with a utility knife or cutout tool.

4. Pry out strips of the finish wall.

5. Pull the drywall-holding nails or screws from studs.

6. Determine the thickness of the nailing strips.

7. Screw in horizontal nailing strips.

8. Secure drywall above and below each strip along each stud.

After the old wall has been stripped out, screw the horizontal nailers to each stud. Also, secure the drywall above and below the cutout strip along each stud, since many of the fasteners holding the drywall may have been removed.

**A drywall cutout tool provides the fastest way to cut out drywall.**

**Use screws to secure backer strips in each stud. Also, secure the drywall above and below the backer strip.**

# Installing Tongue-and-Groove Wainscoting

If you haven't already done this when laying out the backer, start a wainscoting installation by snapping a level chalkline on the wall to indicate the height of the boards. If the floor is reasonably level, all of the boards can be cut to length ahead of time.

The boards should be installed plumb, but if they butt tight to the top edge of an existing baseboard, you may have to cut a slight angle on the bottom end of each one to create a tight fit with an out-of-level baseboard. In this case, you may have to cut each board one at a time, checking the fit as you go. If you do cut an angle on the bottom of each wainscoting board, be aware that the tops won't line up and will instead leave a serrated edge. However, the molding below the chair rail can cover this edge.

## Base first

If the wainscoting will butt into a baseboard, make sure that the baseboard is installed first (see chapter 4 for information on installing baseboard). I like to have a ⅛-in. to ¼-in. reveal where the wainscoting meets the baseboard (depending on the thicknesses of the baseboard and the wainscoting). You can shim out the baseboard so that the wainscoting has an edge on which to sit.

If wainscoting runs all the way to the floor, cut a short piece of ¼-in.-thick plywood or luaun to put on the floor as a temporary spacer to make each board sit off the floor. This gap will later be covered by shoe molding or overlaying baseboard. If the boards butt tight to the top edge of the baseboard, provide a similar gap at the top, just below the chair rail. This makes the boards easier to install and leaves room for them to swell a little in the future, so that they don't buckle or split. The gap at the top will later be

## PRO TIP

*For paint-grade work, outside corners can be butted edge-to-edge, but for stain-grade work, it's better to miter the corner.*

## TRADE SECRET

Make sure that you account for flipping wainscoting boards at corners, so that you always have a tongue facing out. I usually count the number of boards I need for each run and distribute a little pile along each wall, with the top and bottom clearly marked on the back in case I need to cut a slight angle on one end.

## WHAT CAN GO WRONG

If a floor sags, you don't want to mirror the sag on the wall. In that case, establish a level line on the wall, snap a chalkline, and then use that as your bench-mark instead of the floor. Make up for the sag in the baseboard. If a floor has a gentle slope, however, it may be desirable to mimic the slope in the wainscot-ing, so that you end up with par-allel horizontal lines. This is a judgment call that I often need to make in old houses.

covered by cove molding (or a similar detail) running below the chair rail.

## Board layout

I plan the layout so that the wainscoting begins with full widths around windows and doors. I leave partial widths to fall at inside corners where they will be least conspicuous. If there are any outside corners in the room, I start there and work toward the inside corners.

Whenever wainscoting turns a corner—outside or inside—the boards must be flipped end for end, so that you always lead with a tongue. Wainscoting boards can simply butt door and window casing, but an intersecting chair rail should be notched to overlap the casing. Windows with a stool-and-apron design also require extra work. Both of these details are described below (see Wainscoting Meets Window Stool on p. 109).

## Outside corners

For paint-grade work, outside corners can be butted edge to edge. I rip the grooves off two boards (also removing the beveled edge, if there is one) and plane the saw marks off to make a crisp edge. I then test-fit before gluing and nailing the corner closed with 4d (for ⅜-in. material) or 6d (for ¾-in. material) finish nails. I find it works best to nail the two pieces together at the corner, and then attach the assembled corner unit to the wall, sinking the nail in a V-groove or near a bead.

### ✚ SAFETY FIRST

A chopsaw blade tends to grab small pieces of wood and turn them into projectiles. When cutting small pieces, make sure you wear safety glasses. Better yet, wear safety glasses whenever you're working with power tools.

For stain-grade work, it's best to miter the corner.

**1.** Start by ripping a 45-degree angle along the groove edge of two boards.

**2.** Since it can be difficult to prenail a mitered corner, as you would if you were butting the corner, hold the two boards on the corner as close as possible to plumb.

**3.** Tack one board to the wall.

**4.** Apply glue to the corner and tack the second board in place.

**5.** With the two boards secured to the wall, it's easier to push the corner closed and nail it tight.

## Nailing straight runs

When there are no outside corners, I start at an inside corner and face-nail the first board with a single row of nails down the center of the board. For each successive tongue-and-groove board, nail only one edge (the tongue side) of the board. Leave the other free to move—the floating groove slips over the stationary tongue—as the boards expand and contract with changes in temperature and humidity. Secure ⅜-in.-thick boards with 4d finish nails; use 6d or 8d finish nails for thicker material. I check for plumb as I go. To adjust out-of-plumb boards, push the gap between them slightly more open at one end and slightly tighter at the other end. If you have a board with an extreme bow, don't force it over; instead, throw that board in the scrap pile or cut it up for floor blocks.

### WHAT CAN GO WRONG

Walls are rarely as plumb as they should be. Therefore, if you install wainscoting boards plumb, you may wind up meeting an out-of-plumb corner at the end of a run. That is okay. Plane the last board of the run to a slight angle so it fits between the plumb boards and the out-of-plumb corner. And when you turn the corner, start the next run as plumb as possible by cutting the meeting board at the same angle.

Keep a short level handy when you're installing wainscoting, occasionally checking that the boards are plumb.

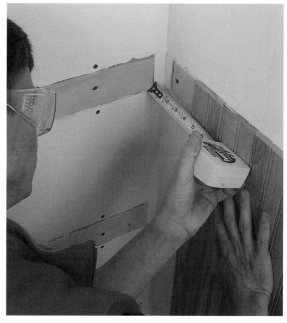

Measure the distance on the wall for the last board in a run.

When nailing, angle the nail toward the panel. If you drive a nail straight in, it's likely to split the fragile tongue. This is certainly one place where I prefer to use a pneumatic nailer. Just make sure that the nail sets all the way. Sometimes the nailer's rubber tip gets in the way when nailing into a corner at the tongue, leaving the head of the nail high. In that case, set the nail by hand.

### Inside corners

Inside corners butt groove to groove. That is, the groove sides of two boards face each other in the corner if you have full-width pieces. Most of the time, however, you have to rip the last board to width, so make sure that you rip off the groove, leaving the tongue for the next board to slip over.

V-groove and bead board look best when identical-width boards face each other in the corner. Therefore, if I rip the last board of a run to width, I also rip the first board of the next run to an identical width so that the adjacent corner boards are mirror images of one another. On the last board, cut off the tongue side (see the

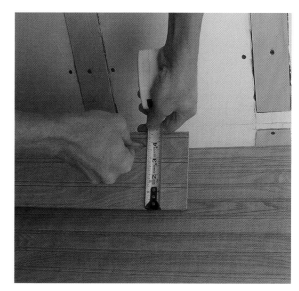

Transfer this distance to the board.

### + SAFETY FIRST

Air nailers are aptly called guns. They fire a nail at a speed that can seriously injure or even kill you. Always stay out of the line of fire and watch where your hands are. If you're working with a helper, make sure he or she is out of the way, too.

**Rip the board to width, cutting off the tongue side.**

photo above). When ripping the facing board, make sure you rip off the groove and leave the tongue to continue the next run. If the walls are out of plumb, this is when you can make up the difference by planing the last board to fit.

# Installing Modified Frame-and-Panel Wainscoting

If you have the time, expertise, and equipment, you can build true frame-and-panel wainscoting that will give a room the look of fine cabinetry. That project is beyond the scope of this book. For the same look at considerably less trouble and expense, you can build a modified frame-and-panel wainscoting system using sheet goods and simple joinery.

## Construction details

I use a frame of 1×4 rails and stiles that I fix to a plywood, MDO, or MDF backer. The backer, which I usually refer to as the "ground" in this application, holds the 1×4 frame in place and

remains exposed to simulate recessed panels. For the look of raised panels, secondary MDF rectangles can be screwed to the ground between the grids created by the frame.

This system works best for painted wainscoting. If you want a natural wood finish, use high-quality veneer plywood as the ground and hardwood rails and stiles. I don't recommend screwing solid-wood fill panels to plywood unless they're narrow—no wider than 8 to 10 in.—and secured with a single center row of screws (countersunk and plugged). Otherwise, the panels are likely to split.

Install the ground directly to the wall studs in place of the typical drywall or plaster wall covering. Lightly tack each panel in place with a few 4d finish nails, and then make layout marks on the plywood that describe where the rails and overlay panels will go. Permanently secure the ground with 2-in. wood screws, hiding the screw heads below the rails (and overlay panels, if there are any) wherever possible.

I like to use clear poplar for the frame because it takes paint well, is relatively stable, and is often less expensive than clear pine. Typically, I install a

**Plywood backer serves as the surface for a recessed panel over which rails and stiles are applied.**

103

## PRO TIP

*Chair rail should include a small finish trim below to hide the expansion gap, which is particularly important when wainscoting fits tightly to the baseboard.*

## TRADE SECRET

If I need to bury a screw head to get maximum length but don't need to worry about finishing the hole, I use Fuller® countersink bits (see Resources on p. 167). They're inexpensive (about $5 each), use standard twist drill bits, and are easy to adjust for length. I prefer them to most other inexpensive countersinks because the cutterhead stays sharp longer and doesn't tear the grain as severely.

## WHAT CAN GO WRONG

When fastening rails and stiles, or solid panels to a plywood ground, don't use pairs of screws. This prevents the rail from expanding and contracting with seasonal changes in humidity, which may cause the board to crack. A single row of screws along the center of the board is enough to hold it in place.

Level the prebuilt panel.

Screw securely to each stud.

Apply drywall above the panel.

1×4 rail along the top, a 1×6 or 1×8 baseboard along the bottom, and 1×4 uprights in between. As always, select dry material to avoid cupping and warping, and prime the back side of the rail stock before cutting it to length and securing it to the ground.

Secure the rails with 1⅝-in. wood screws, countersinking the holes below the surface. Lay out the screws in a single line running down the middle of each rail, which allows the rails to move a little as they shrink and swell. In addition to screws, install biscuits at the ends of stiles where they intersect with the top and bottom horizontal boards. This helps align the surface of the boards and provides a tight joint that's less likely to crack the paint. I join the plates as I install the frame.

## Prebuilt wainscoting panels

Although I typically build wainscoting panels in place, it's also possible to prebuild them in full-length sheets that can be attached to the wall studs in one piece. The panels can then be built in the shop under controlled conditions. On site, the panels are leveled and screwed to the framing to quickly finish a large wall area at one time (see the photos above).

# Installing Chair Rail

Chair rail, which caps wainscoting, can be its own detail. In rooms broken up by a lot of windows, chair rail can provide plenty of extra interior detail where full wainscoting would be overkill. By itself, chair rail produces an immediate and powerful effect of dividing the wall height into unequal portions. Painting the two portions in different, complementary colors can enhance the effect.

When used in conjunction with wainscoting, chair rail should include a small finish trim below to hide the expansion gap, which is particularly important when wainscoting fits tightly to the baseboard. Although there are a number of possible chair rail configurations (see the drawing on p. 93), I typically cut a simple chair rail from 5/4 stock, ripping it to 1½ in. wide and rounding over the outside edge with a ¾-in. roundover bit.

When I'm working with stain-grade chair rail, I usually nail it in place with stout 10d to 12d finish nails, angling them from the top down toward each stud in the wall. Predrill for these nails, then set the nail heads. A finish nail is easier to hide than a screw and plug.

As with other horizontal trim, such as baseboard and crown, the outside corners should be mitered and the inside corners coped. Similarly, linear runs of both the chair rail and the underlying scotia should be joined by a scarf joint.

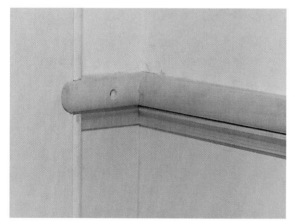

A chair rail ends gracefully as it overlaps the door casing. Note the scotia ends at the edge of the casing.

This is an example of a properly fitted cope.

If possible, rout the edges before ripping chair rail. When necessary, brace narrow stock against blocks clamped to a benchtop, as shown.

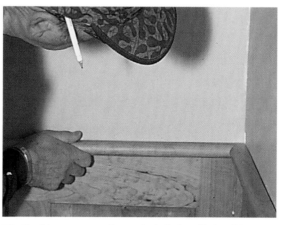

The inside corners of rounded chair rail should be coped.

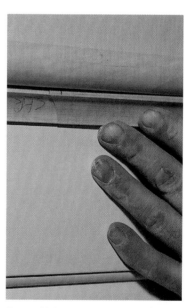

A scarf joint is used to join lengths of trim.

## TRADE SECRET

When I want to install a wood plug to hide a screw head, I use a Forstner-style bit to drill about 3/8 in. deep (the approximate thickness of the bit's cutterhead). Forstner bits cut a clean hole that perfectly matches the diameter of standard dowel stock or plug cutters.

## IN DETAIL

Use a nail set, or nail punch, to countersink the head of a finish nail below the surface of woodwork. These typically come in sizes ranging from 1/16 in. to 5/32 in. for use with different size nails. I usually carry just two nail sets: 1/16 in. (often specified as 2/32 in.) for most finish nails and 1/8 in. for larger finish nails. For the occasional brad, I do have a 1/13-in. nail set in my toolbox, but I prefer to use a brad setter rather than a hammer and nail set. When using nail sets, always wear safety glasses.

## Scribing chair rail

Most of the time, chair rail must be scribed to the fluctuating surface of the wall. Besides uneven wall surfaces, outside corners usually have a buildup of mud near the corner bead, so they also require scribing. The problem is that scribing the rail can throw it out of parallel with the face of the wainscoting, so a miter at the outside corner may not be a perfect 45-degree angle. I approach the task this way:

**1.** Cut the rail on one side of the corner to length with a 45-degree miter on the end.

**2.** Holding the rail on the top of the wainscoting, scribe the back edge of the rail to match the wall and plane the rail to the line with a block plane.

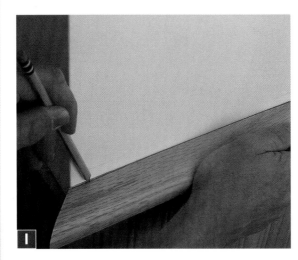

**3.** Predrill for a 10d finish nail and tack the scribed rail in place, nailing at an angle into a stud.

**4.** Hold the next rail to the wall, pushing it hard to the outside corner and parallel to the face of the wainscoting. Mark the matching miter angle, tracing underneath the second rail and against the first rail.

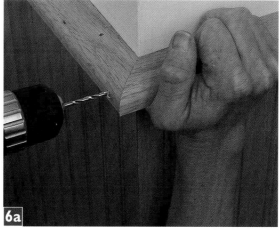

**5.** After cutting the scribed angle, scribe the rail to the wall, planing the back edge of the chair rail until it has a tight fit.

**6.** Nail the miter closed with opposing finish nails. Depending on how much is planed off the scribed edge, the outside corner may have to be fine-tuned with a block plane.

**7.** Finish up the chair rail by installing trim molding.

107

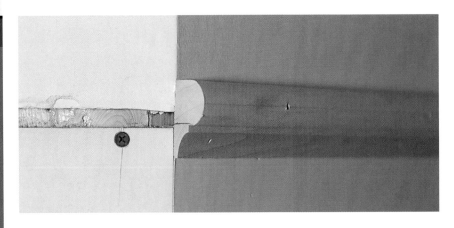

The intersections of different kinds of trim require planning and careful joinery.

## IN DETAIL

If you're serious about trim-work, eventually you'll want to own a pneumatic trim nailer; at least 25 ft. of hose; and a small, portable compressor. With a single pull of the trigger, a finish nailer can fire a 2-in.-long nail though hardwood without splitting the ends, leaving hammer marks, or requiring an extra step to set the nails.

## TRADE SECRET

On long wall sections, it may be necessary to scarf together short sections of various chair rail components. Stagger the scarf joints in the mail rail and the finish moldings to make them less conspicuous.

### Finishing details

The details at windows and doors take the most time and attention. The wainscoting can simply butt the casing. The chair rail, however, should be notched so it overlaps, and the scotia below should butt the casing. If the casing has a bead or a back band, notch the casing as well as the chair rail.

Around square-edged casings, you can simply notch the rail so that it overlaps the same distance as the thickness of the rail. However, when the casing has a back band or a beaded edge, you also need to notch the casing—in effect, weaving the joint together. Here's how:

**1.** Measure the dimensions of the notch by holding the rail on the wainscoting and marking the edges with a sharp pencil.

**2.** Cut the notch in the casing. Here, carpenter Craig Tougas uses a Japanese pull saw, keeping it square to the edge of the casing by resting it on a block of wood.

**3.** Once the sides have been cut, finish off the notch in the casing with a chisel. Test-fit the notch by holding the rail in place. Mark the depth of the casing on the end grain of the rail.

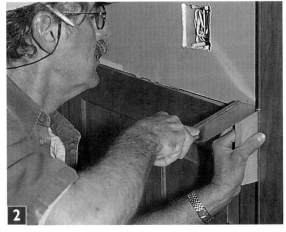

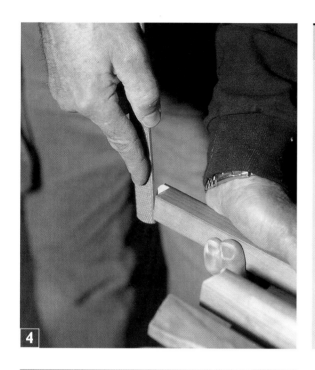

**4**

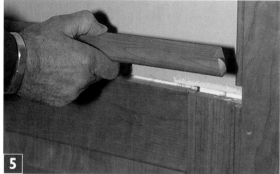

**5**

## Caulk for Paint-Grade Trim

For caulking interior woodwork, choose an acrylic latex sealant. These nontoxic formulations are easy to tool with a wet finger, clean up with water, have good resistance to ultraviolet (UV) light, and take paint well.

Because they contain water, latex acrylics shrink as they cure—up to 30% in volume. For most small gaps (less than 1/16 in.), the shrinkage isn't too noticeable. But for larger gaps, you may have to add another layer of caulk before painting.

There are many grades of latex caulks, including polyvinyl acetates, vinyl acrylics, and 100% acrylics. Most of these formulations are "siliconized," which means a small amount of silane (a form of silicone) has been added to promote adhesion. For top performance, look for a product with the ASTM C 920 Class 25 designation. This standard indicates that the sealant can move up to 25% of its volume without splitting or peeling away.

**4.** Notch the rail with the same pull saw, using a rasp to smooth the edges to a crisp, square cut.

**5.** When finished, check the final fit.

Stools and aprons on windows require even more work. If the window is already trimmed, pull off the apron and replace it with a rabbeted apron. If the window hasn't been trimmed, use a scrap of casing stock to measure the distance from the jamb (including the reveal) to the outside edge of the casing, mark the wall, and run the wainscoting to that line. If it doesn't break on a full-width piece at the outside casing line, notch the wainscoting underneath the window. The notch then comes in under the stool or under the apron, depending on how you handle the apron.

**Wainscoting Meets Window Stool**

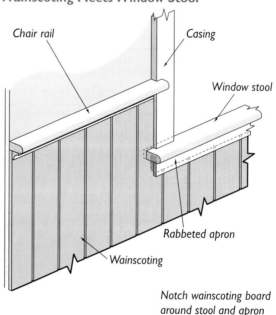

Chair rail

Casing

Window stool

Rabbeted apron

Wainscoting

*Notch wainscoting board around stool and apron*

**Rabbeted Apron**

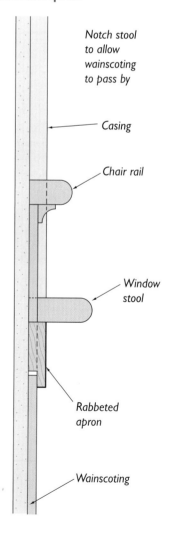

*Notch stool to allow wainscoting to pass by*

Casing

Chair rail

Window stool

Rabbeted apron

Wainscoting

Install the stool after the wainscoting is in place, so that the horns overlap the wainscoting, as shown. I've seen many carpenters cut a wide stool, and then install the apron on top of the wainscoting. But I prefer to bring in the notch under a rabbet on the apron. This way, all of the window stools in the house are the same width and all of the window trim—casings and aprons—lie on the same plane.

# Materials for

# CHAPTER SEVEN
# Built-Ins

**1** Plywood, p. 112

**2** Composite Panel Stock, p. 117

Usually the term "built-in" refers to a cabinet or shelving that is literally built in (and permanently fixed) to the house. I prefer to think of built-ins as a subclass of interior trim carpentry that includes slightly larger projects, such as wainscoting and fireplace surrounds. While these projects are built with the moldings and finish lumber described in detail in chapter 2, they also depend heavily on long, wide panels, or *sheet goods*.

Plywood is the most familiar form of sheet good. Today, the options for panel stock range far beyond plywood and include composite materials, such as particleboard and medium-density fiberboard. Add plastic laminate, melamine, and hardwood veneer, and you have a wide variety of panel stock from which to choose. In this chapter, I'll describe the various panels and materials that I use for built-ins, where and how I use them, and how to buy them.

## PRO TIP

*If you're storing panel stock flat on a concrete slab, lay down plastic sheeting first. Otherwise, the plywood will absorb moisture from the concrete.*

## IN DETAIL

The quality of the veneer on both faces depends on the grade of the plywood; therefore, a grade designation is assigned to each face. For example, plywood designated AA has two A-grade faces. This is the best quality available, but it's usually a special-order item. A sheet marked AC has one A-grade face and one C-grade face.

## WHAT CAN GO WRONG

The biggest drawback to luaun is that it has heavy extractives in the wood, which usually bleed through water-based paint. To avoid this, use a stain-blocking primer first.

## TRADE SECRET

Regardless of where you buy panel stock, make sure that you buy enough, especially if you are planning to apply a natural finish and are matching the grain. Even if you don't buy numbered runs, hardwood is typically sold in consistent batches. If you come back in a week for that extra sheet, chances are high that both the color and the grain will vary significantly from the sheets you purchased earlier.

# Plywood

Plywood is always my first choice for built-ins. When quality is more of a consideration than price, I almost always choose ¾-in.-thick veneer-core plywood. For paint-grade built-ins, I opt for birch-faced plywood, which takes paint exceptionally well. For natural finish cabinets, I usually choose maple-, oak-, or cherry-face veneer—whatever fits best.

As the name suggests, plywood is made from numerous plies of thin wood, called "veneer." These thin sheets of wood are peeled from a tree round. Both the core—the inner layers—and the face—the outer layers—are considered veneer, though thickness and quality vary. The best wood is used for the face, while the core is made from the plainest, even slightly flawed wood. The core layers are laminated in alternating directions, so the grain of one layer runs crosswise to the other. Because of these alternating layers, plywood

Hardwood plywood often has the same core—laminations of poplar wood veneer—with different hardwood face veneers, including cherry (top), oak (center), and birch (bottom).

Luaun plywood is a suitable material for cabinet backs. It's available with one side faced with hardwood veneer, which is appropriate for the inside face of cabinets. By itself, luaun does not accept paint or stain well.

is much stronger and more stable than solid wood.

## Panel thickness

Like most sheet goods, plywood is commonly available in thicknesses ranging from ¼ in. to 1⅛ in. I build nearly all built-in cases from ¾-in.-thick material. This thickness provides ample strength for joining the edges of pieces together and tends to stay very flat. To save money or to create a lightweight, portable cabinet, ½-in. stock can be used for short shelving runs. However, this thinner material presents problems when joining pieces for built-ins. For cabinet backs and finished end panels, I typically use ¼-in. stock. For exposed cabinet backs (such as those on an island cabinet), I prefer to use ⅜-in.-thick plywood. For the backs of bookshelves and built-in cabinets, I generally use nominal ¼-in. luaun secured with wood glue and 1-in. ring-shank nails.

Luaun panels actually measure 5.5mm thick (¼ in. = 6.35mm). The panels are measured in the metric system because they are imported from the Pacific Rim, where luaun (a type of fast-growing mahogany) is harvested. You can buy luaun just about anywhere, and it is the most affordable plywood option. Three face-grades are available: BB, CC, and OVL (overlay). Knot holes ¾ in. in diameter are permitted in the face of the worst grade (OVL) if they are puttied. The grade is often stamped on the edge.

Luaun is available with a hardwood veneer on one or both surfaces for naturally finished cabinet backs. For exposed backs, such as those on an island shelf, I prefer to use ⅜-in.-thick hardwood plywood. The reason for this is simply strength. I once watched a fast-moving 12-year-old boy put his foot through a hardwood-faced luaun cabinet back. This was an accident, but it still proved very difficult to repair.

## Plywood dimensions

Plywood is always designated first by the short-grain dimension, and then by the long-grain dimension. Mostly, you will see 4×8 material in which the grain runs along the 8-ft. dimension. Occasionally, you may find 8×4 panels. Called *counterfront plywood* (and typically a special-order item), these panels have face grain running along the width of the sheet, providing evenly matched grain for cabinet fronts.

Plywood is also offered in longer lengths— 10- and 12-ft.-long sheets—as a special-order item. As you'd expect, you will pay a high premium for these nonstandard materials. But for exceptionally tall or long runs of cabinetry, these special configurations are well worth the cost.

## Core options

Plywood is a two-part material, with face veneers on each side sandwiching the core. The core in

### Shoot Board for Cutting Plywood

A "shoot board" first serves as a straightedge to guide the saw. But just as important, it also holds down the grain along the cut so that the blade doesn't tear out on cross-grain veneer. I made my shoot board with two strips of ¾-in.-thick material, using one piece for the fence and one piece for the table. The table is the width of the shoe on my saw, which rides on the table. I use a pair of spring clamps to secure the outside edge of the board's table to the cutline on the plywood. With veneer that splinters easily, such as oak, I first score the cut along the edge of the board. I replace the table on the shoot board occasionally because different blade thicknesses chew up the edge near the blade.

most hardwood plywood is typically made from lower-grade laminations of hardwood than the face veneers. Poplar, birch, and mahogany are common woods used, either as a *lumber core* or as a *veneer core*.

**Lumber-core plywood.** A special-order item from most lumberyards, lumber-core plywood is strong and stable, and it holds screws and

Most plywood is made from individual laminations of veneer layers (top). Lumber-core plywood (bottom) is a special-order material that holds edge screws better, making it an excellent option for furniture and cabinets.

## PRO TIP

*When choosing panel stock for painted cabinets, consider using MDF, which offers the best value in sheet goods.*

## WHAT CAN GO WRONG

Use fir and other softwood plywoods for framing construction projects and as substrate for flooring and plastic laminates. While AC-grade fir plywood often looks smooth and even, it doesn't make a very good paint substrate. The grain is often raised by paint and the face usually absorbs paint unevenly. The raised grain and patches in the face veneer usually telegraph through any finish. Look closely at the core—it often has fairly large voids. These make the edges unstable for joining and may telegraph through paint.

## TOOLS AND MATERIALS

The price of hardwood plywood varies enormously, depending on the type and species of the core and the face. It's worth shopping around—call the more popular lumberyards and home centers as well as the local hardwood dealers. (Look under both Lumber and Hardwoods in the phone book.) In my experience, smaller hardwood dealers offer comparable or slightly better pricing and often much better quality than larger chain outlets.

other hardware as well as solid wood does. But it's pricey, averaging about twice as much as comparable veneer-core stock. Lumber-core material should have no voids. However, some imported varieties made with luaun as the core stock have voids caused by core-piece shrinkage. If possible, steer clear of these materials.

**Veneer-core plywood.** This is the plywood you'll usually find at a lumberyard. It, too, is strong and stable, and holds screws reasonably well. However, whenever you drive screws into the edge, always predrill to avoid separating the plies.

Veneer-core plywood is made in a variety of ways to serve different structural and aesthetic purposes. For example, the core veneers in aircraft plywood are laminated at 45 degrees to each other, making one of the strongest wood-based materials available. Marine plywood is made with waterproof glue cured under pressure so that it

### + SAFETY FIRST

Heavy panel stock, such as ¾-in. plywood and MDO, can be a real back breaker. If possible, get help when moving sheets. If you must carry sheets by yourself, balance the sheet from the center, holding the bottom and top edges, as shown. In this case, the shop has been sensibly arranged, with the distance between the storage area and the table saw short enough so that the panels are always supported.

**For cabinets, look for high-quality material with at least seven veneer plies (bottom). Five-ply material tends to have more voids in the core (shown on the edge of the top piece of plywood).**

won't delaminate, even under the most extreme conditions of moisture and temperature.

Typical furniture- and cabinet-grade plywood has five or seven plies and sometimes more. Generally, more plies indicate better quality.

Five-ply material may have some voids in the core, which can present a problem with shelf pins. If a hole for a shelf pin falls on a void, the pin is likely to sag or fall out. Higher-grade seven-ply material tends to have fewer voids and is less likely to split when driving fasteners into the edge.

## Face options

All cabinet-grade plywoods have either rotary-cut or flat-cut face veneers.

**Rotary-cut veneer** is peeled from a spinning log in a long, continuous sheet. This type of veneer is commonly used for softwood plywood, but it tends to be a poor choice for cabinet-grade material, because of its strong, unappealing grain patterns and defects caused by trying to make the curved sheet lie flat.

Softwood plywood is nearly always made from rotary-cut veneer, which allows the face to be one continuous piece. However, stresses in the peeled veneer often crack.

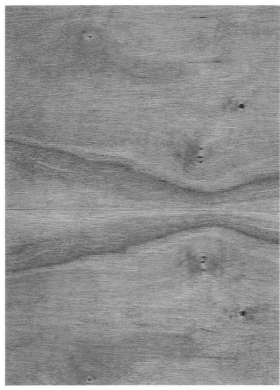

True book-matched veneer has an interesting matching grain pattern and the matching flitches absorb stain and clear finish evenly.

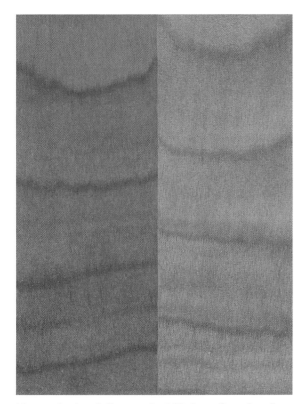

Unmatched and slip-matched flitches often absorb finish differently, creating an unsightly striped effect.

**Flat-cut veneer** is made by running the log through a stationary knife, which slices off narrow sheets. These sheets (called "flitches") are joined side by side on the panel, making for a more expensive, but also a more desirable, cabinet-grade panel stock.

The individual flitches on the face of plywood typically vary in size from sheet to sheet, and the width and number of flitches used in each sheet affect the appearance. For naturally finished cabinets, it's often worth paying close attention to how these veneers are joined together. For example, when I lay out a series of cabinet doors, I locate the seam of the book-match and measure from it in equal directions. The seam can be placed either in the center or in between doors, depending on how many different cabinet doors need to match. I also try to choose upper and lower cabinet doors from the same vertical sheet, so that the grain runs continuous from top to bottom.

### Rotary-Cut Veneer vs. Flat-Cut Veneer

*Rotary-cut veneer is cut with the log spinning into the knife to create one continuous peel. Flat-cut veneer is cut with the log moving straight across the knife to produce many narrow sheets called "flitches."*

**Rotary Cut**

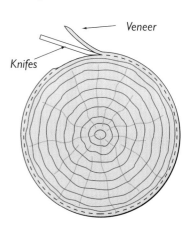

**Flat Cut**

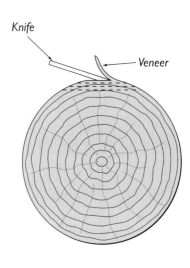

**PRO TIP**

*Other options for dressing up built-ins include special thick veneers and veneers with paper backing that are applied with contact cement.*

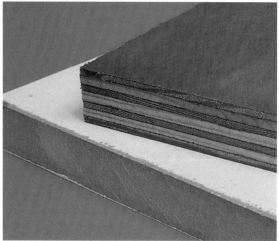

MDF vs. MDO: Although the acronyms are similar, the materials are not. Medium-density fiberboard (MDF) is a composite material made from fine wood fibers (bottom); medium-density overlay (MDO) is a plywood product with a paper face overlay (top).

## Storing Plywood

The best way to store sheet goods so that they don't twist or bow is to keep them perfectly vertical on edge or absolutely flat and continuously supported. Full sheets of panel stock take up a lot of room, so it's often easier to store them on edge. Place them on short 2×4 blocks to keep the edges away from moisture and keep the panels as vertical as possible. If they are stored at an angle, the weight of the panels can cause them to bow.

Panels may be stored horizontally, but they must be fully supported so that they don't sag.

A better solution is to store panels as close as possible to vertical to prevent them from sagging.

Long runs of cabinetry may be built from several sheets of plywood. You can buy up to twelve sheets of flitch-matched plywood to make the run consistent. Such sheets are marked consecutively as 1/12, 2/12, 3/12, and so on. However, they require placing a special order with a hardwood dealer—a luxury I don't always have. Usually, I lean full sheets of plywood against the wall and rearrange them until I obtain the most pleasing composition possible. Be sure to number the doors and keep track of each piece as you work.

Whenever possible, I use true book-matched veneers. Often, slip-matched material looks convincingly like similar grain, but the wood absorbs finish quite differently, creating a distinct—and rarely acceptable—striped effect once the panels are finished.

**Plywood grades.** Typically, the grade of plywood is specified by the quality of the face veneers. The two faces are not necessarily of the

Today's choices in panel materials include exotic sandwiches. From top to bottom: plywood with an MDF core, plywood with a traditional veneer core, particleboard, and plywood with a combined MDF/veneer core.

same type or quality of veneer. An A face—the best quality available—has no knots, splits, or other defects. B and C are of lesser quality but often highly acceptable for painted built-ins, though plan on putting a C face inside the cabinet.

# Composite Panel Stock

While I think hardwood plywood is the easiest and most reliable material to use for built-ins, it's certainly not the most affordable or necessarily the best. In the past decade, medium-density fiberboard (MDF) has improved tremendously, and perhaps more than any other product, offers the best value. Cabinet-grade particleboard has also improved and is very affordable, though I think it is the least desirable material available for built-ins.

Both MDF and particleboard are used as the substrate for panels with high-tech finishes, such

**The unsealed edges of MDF absorb paint and moisture and are generally unstable (above). (The MDF below is painted.)**

## Plywood Face Veneers

*Flitch-matched* panels use veneer sliced from the same log. Individual flitches on a 4×8 panel are commonly arranged in two ways: book-matched and slip-matched.

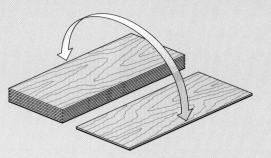

### Book-Matched Veneer

To create book-matched veneer, a plywood manufacturer turns over the top flitch from a stack of consecutively sliced flitches like the page of a book, then joins these two pieces side by side. This way, the grain on one surface mirrors the grain of the next. Such pairs of flitches make up the face of a sheet of plywood.

### Slip-Matched Veneer

This veneer is made by joining consecutive flitches without turning them over; in this case, the grain pattern runs parallel rather than as a mirror image. This grain pattern repetition works best with especially wild and distinctive grain. However, slip-matched veneer makes it difficult to achieve a sense of symmetry in the grain design across a cabinet face. For that reason, I prefer using book-matched material, which is, fortunately, the more common of the two.

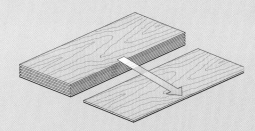

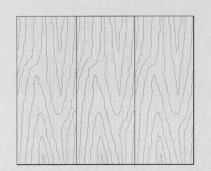

## PRO TIP

*MDF absorbs moisture and can crumble. Avoid using it in damp locations, such as bathrooms.*

## IN DETAIL

Plastic laminate comes in two basic grades: horizontal and post forming. Horizontal grade, which is about 1/16 in. thick, is what you should buy to apply to a substrate. Post-forming material is much thinner and is used for wrapping over the rounded edges and molded backsplash of a post-formed countertop.

## WHAT CAN GO WRONG

Nothing will make your heart sink as fast as a broken corner or a nail blowout in a pristine melamine surface. However, such an unsightly scar can be repaired using epoxy with a universal color mix. Both ingredients can be purchased from a paint store or a large hardware outlet that sells paint. Blend the epoxy and the color, and then apply the mixture with a sharp chisel. After it cures, pare it smooth. What you can't do is sand and paint this material. No matter how much you try, you will never be able to match the texture of the melamine surface.

Although it is more commonly used for exterior signs and panels, **MDO** takes paint better than any other panel product and doesn't delaminate in high humidity—features that qualify it for cabinet material in special applications, such as patio barbecue stations, poolside utility cabinets, and steam-room vanities.

Both MDF (left) and particleboard (right) are composite materials made from shredded wood fibers and adhesive bound under heat and pressure. However, because the flakes in particleboard are so much bigger than the fine fibers in MDF, when particleboard chips or has a nail blowout, the results are much more disastrous.

as melamine and plastic laminates. These products offer outstanding performance and come in an enormous range of colors and styles, making them the products of choice for some of the finest pre-fabricated cabinetry on the market.

## MDF

MDF is made from very fine wood fibers that are mixed with a urea-formaldehyde–based adhesive, and then compressed under heat and pressure. The result is a stable material with a super-smooth surface that takes paint exceptionally well. It also makes a good substrate for plastic laminate and veneer.

MDF usually comes in sheets measuring 49×97, which provides plenty of area for ripping full 12-in. or 24-in.-wide pieces. Currently a plain, 3/4-in.-thick sheet sells for about a third of the price of hardwood plywood.

Although I prefer using plywood, MDF is so much more affordable that I now use it more than almost any other material for paint-grade built-ins, particularly bookcases. However, it has some definite drawbacks that you must be aware of before using it. The most obvious one is weight. A full sheet of 3/4-in.-thick MDF weighs close to 100 lb. (compare that to a sheet of hardwood ply-wood, which weighs about 70 lb.), so you defi-nitely need help handling full sheets. Although the homogenous thickness looks uniform, the edges drink paint, so they must be banded. And just as

## + SAFETY FIRST

When cut, the tiny fibers of MDF pulverize into a super-fine dust that defy dust collection systems. The formaldehyde-based adhesive used to bind the fibers is also irritating and possibly carcinogenic. You must wear a dust mask when cutting this material; plan to spend a lot more time vacuuming the dust as well.

the edges drink paint, they also absorb moisture, making the edges swell and, in extreme cases, crumble. I do not recommend using MDF in bathrooms or any location where it may get wet.

MDF is available in a wide variety of hardwood-face veneers (for example, Georgia-Pacific® Fiber-Ply). Although this material resembles plywood, it is much more delicate. Because the surface of the MDF substrate is so smooth, manufacturers are able to use an ultra-thin face veneer that is much easier to accidentally sand through, and the edges are still quite fragile.

## MDO

Not to be confused with MDF, medium-density *overlay* (MDO) is an exterior-grade plywood covered by a resin-treated paper. This material has a glass-smooth surface especially made for paint. It is a popular product for signs and makes a great material for built-ins, combining the advantages of veneer-core hardwood plywood with the best paint surface available. However, MDO is expensive, typically costing about 10% more than comparable hardwood plywood. Nevertheless, you should seriously considered it for built-ins in damp locations because it is made of an exterior-grade plywood that won't delaminate when periodically exposed to water.

## Particleboard

Particleboard has been around for a long time, and modern versions are vastly superior to those made 20 or more years ago. For starters, the chip size, and therefore the panel density, varies across the thickness. Larger chips fall to the core and finer chips fall on each surface. This makes for a stable product with a smooth surface for wood veneer or plastic laminate. Particleboard is a common substrate for melamine and plastic-laminate panels. It also comes preveneered in a variety of hardwoods, mostly birch and oak.

### Nail Blowouts

Joining thin panel stock presents a number of difficulties. The most common is nail blowouts. Unless a nail is centered within the edge of the panel and driven perfectly straight, it can easily blow out the sides. This happens most frequently when you use a pneumatic nailer, but it is also likely to occur when you use a hammer or even drive screws.

To fix a blowout, use end nippers to clip off the nail below the surface. If you can't clip the nail below the surface, use a nail set to drive the cut end into the panel. Use a sharp chisel to pare down the bulged material around the blowout. Last, fill the hole with epoxy wood filler. Here, too, I use a sharp chisel to apply the filler and leave a clean surface that requires minimal sanding.

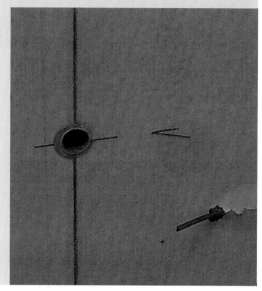

Despite recent improvements, I am not a big fan of particleboard. Although it takes paint well, paint may raise the particles slightly, so painted particleboard is not as smooth as painted MDF. Also, particleboard is just as heavy as MDF and the corners are just as fragile. But because the particles are so much bigger than those in MDF are, when particleboard chips or has a nail blowout, the results are much more disastrous. And though the faces of a panel have a better screw-holding capacity, the edges do not. I generally steer clear of particleboard except when I'm using melamine and plastic laminate.

## Specialty surfaces

Both particleboard and MDF are available with special finishes made of resin-saturated industrial papers. The most common and versatile ones for built-ins are melamine and plastic laminates.

## PRO TIP

*Melamine panels offer a tough, smooth surface, but because the brittle surface chips easily, they require careful joinery.*

## IN DETAIL

MDF cuts cleanly with no tearout. But the corners become very fragile and can easily chip or dent when handling cut-out panel pieces. Plan to use biscuits, glue, and nails for joining MDF. Screws hold poorly in the face and especially badly in the edges.

## IN DETAIL

Melamine advantages:
- Prefinished.
- Tough surface resists water.
- Shelving precut and predrilled.
- Edges can be finished with iron-on tape.

**Melamine-faced particleboard provides a prefinished panel material. But the corners chip easily, so it requires ultra-sharp tools and the utmost care.**

**Glue doesn't stick to melamine, so the edges must be rabbeted.**

**Melamine.** Although it is often referred to as a separate panel material, melamine is a clear resin used to create a panel surface. True melamine panels use industrial paper with a basis weight of 80 to 140 grams. The paper comes in a range of colors (white is the most popular) and decorative surfaces (ones that mimic wood grain are the most common). The paper is thoroughly saturated with resin, laid over the substrate (particleboard, MDF, or plywood), and fed into a press, which applies heat and pressure at about 350 psi. During this process, the resin penetrates the substrate fibers, permanently fusing the surface to the substrate. The result is a surface that's thinner than but similar to plastic laminate.

The biggest advantage of melamine panels is that the finish is already done, and it's smoother and more consistent than just about any hand-applied paint. The surface is tough, wears well, and resists water. Melamine is also available predrilled and preripped for shelf stock, which offers a huge laborsaving advantage.

The down side of melamine is that it requires more elaborate joinery and the utmost precision. Glue doesn't stick to melamine, so the joints must be rabbeted. And the brittle surface chips easily, requiring extra care and exceptionally sharp tools. The edges are usually taped with an iron-on polyester ribbon.

**Plastic laminates.** Plastic laminates are made from multiple layers of paper saturated with phenolic resin, resulting in a much more rugged surface than melamine. Typically, plastic laminate used for cabinet faces and countertops

has three parts: 1) a core of layered paper saturated with phenolic resin and bonded under
high pressure (about 1400 psi), 2) a colored or
patterned paper, and 3) a clear melamine top
coating. This process produces a ¹⁄₁₆-in.-thick
sheet of plastic that comes in lengths of up to
12 ft. Plastic laminate can be purchased separately from the substrate. (For more on applying
plastic laminate, see In Detail on p. 128.)

Sometimes referred to as high-pressure
laminate or HPL, plastic laminate comes in
an astounding variety of colors and patterns.
Between the two major manufacturers—
Formica® and Wilsonart®—you can match
just about any color and style you could want,
including faux wood, stone, leather, and fabric.
Standard stock always has a dark-colored resin
core that shows a brown line along the edge.
However, both manufacturers also offer laminates that are the same color all the way though,
making the seams less conspicuous.

**Layers in Plastic Laminate**

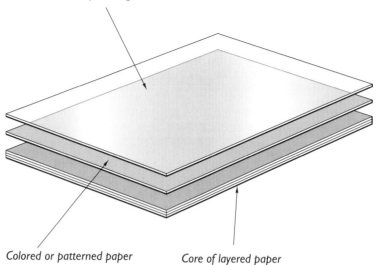

*Clear melamine top coating*

*Colored or patterned paper*

*Core of layered paper*

*Plastic laminate used for cabinet faces and countertops
typically has three parts: 1) a core of layered paper
saturated with phenolic resin; 2) a colored or patterned
paper; and 3) a clear melamine top coating.*

**Plastic laminate is a versatile
material that can be used in
a number of applications. Here,
a post-formed countertop—
a particleboard underlayment
topped with a plastic laminate
surface—is being prepared for
a kitchen installation.**

# Basic Bookshelves

# CHAPTER EIGHT

From simple wall brackets that hold up plain boards to elaborate pieces of furniture, bookshelves come in a wide variety of designs. In this chapter, I'll describe my simple, adaptable method of building adjustable shelves with hardwood face banding. This flexible design works well for floor-to-ceiling installations that cover either all or part of a wall, and it can wrap around a corner, too.

The heart of my system is a simple plywood box, the same format upon which kitchen cabinets and bathroom vanities are built. All of these are designed as "units" of plywood boxes. Each one is built to different proportions with different combinations of shelves, doors, drawers, and countertops. You can assemble shelf units, kitchen cabinets, vanities, and other built-ins by following the same basic methods described here.

## IN DETAIL

One way to integrate the design of a built-in is to repeat architectural features of the house. For example, try to match the proportions of rectangular cases to doors and windows. Consider the orientation of shapes. If the windows run horizontally, align the cases horizontally. If they run vertically, align the cases vertically.

## TRADE SECRET

Here are some design tricks to keep in mind:
- Think in units: Create multiple bookcases and join them like building blocks.
- Determine the kind of shelving you want. Adjustable shelves provide maximum flexibility, while fixed shelves help stiffen a case. Plan on at least one fixed shelf in tall cases (70 in. or taller) aligned with horizontal features in the room.
- Match trim details. Design the face frame, shelf aprons, and fascia to reflect other woodwork in the house.
- Accommodate furniture. Bookshelves take up wall space, so consider how freestanding furniture will also fit inside the room.

# Bookshelf Design

**G**ood shelving not only complements its surroundings, but it also has structural integrity. No matter how pleasing the proportions of a built-in are, no matter how beautiful the finish on the woodwork is, your work will be a disappointment if the shelves sag or the cases rack when loaded with books. So when planning bookcases, always account for both appearance and strength.

## Design strategies

First, let's focus on what the shelves look like as an interior design element of the house. Here are some strategies to keep in mind when integrating built-in bookcases with the rest of the house:

**Think in units.** A wall of shelving is built as a series of vertical boxes constructed from plywood and solid-wood edging. Since the shelves are built in box units, design the shelving as a series of rectangles.

### Bookshelf Planning

*Bookshelves are built in box units, so design the shelving as a series of rectangles, paying attention to both the vertical and the horizontal orientation of architectural features in the room.*

**1. Vertical Formats**

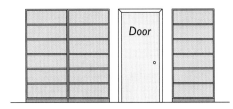
Door

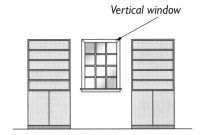
Vertical window

**2. Horizontal Formats**

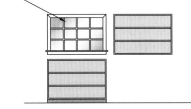
Horizontal Window

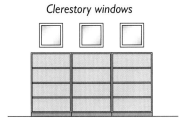
Clerestory windows

**3. Combined Formats**

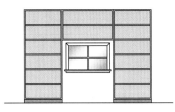

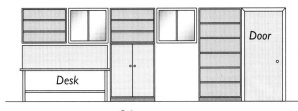
Casement windows
Door
Desk
Cabinet

A simple but elegant built-in shelf makes efficient use of space and visually ties in with the room's wood trim.

**Complement the room.** Any built-in cabinet should fit in as a natural part of the room. Unlike furniture that will come and go, built-in bookcases will become part of the house—an interior feature as lasting as the window trim and baseboards. For this reason, I typically steer customers away from designing built-ins to match freestanding pieces of furniture that will leave when the occupants move.

Often, bookcases run from floor to ceiling to maximize shelf space. In this case, it's not the overall box dimensions but the placement of the verticals and fixed shelves that reflect the room's elements. For example, I typically plan to have at least one fixed shelf in tall shelf cases (70 in. or taller) to stiffen the sides. Depending on the room's features, I put this fixed shelf at a height that matches the important horizontal elements in the room.

In addition to matching architectural shapes, try to match trim details. Back bands, fluting, and reveals on window and door casing can be mirrored in the face frame of the built-in. Even an arch-top window in the room may call for an arch cut into the head of the bookcase. However, if you're ever in doubt about what will look good, err on the side of simplicity.

**Accommodate furniture.** Bookcases must accommodate other furniture in the room. Often, this leads to a battle for wall space. If an entire wall is taken over by shelving, you won't be able to put the couch along that wall. Do you have the floor space for a couch, reading chairs, tables, and other freestanding furniture to occupy the center of the room? Or do some of these pieces need to be pushed against the wall? If so, can the bookcases be built in units that fit around the furniture?

When planning all of this, don't forget that your house is an investment, both for yourself and for your family. Always weigh what will work for you and your particular needs against and what will work in a general way for future occupants. The best design is a thoughtful balance of these considerations.

Custom built-in shelves can accomplish many tasks. Here they serve not only as shelving, but as a plant stand and radiator cover as well.

**PRO TIP**

*A solid-wood 1×2 apron banded to a shelf acts as a beam, almost doubling the load-carrying capacity of the shelf.*

## IN DETAIL

I try to place fixed shelves at the same height as window sills or at the height of windows and doors, depending on which features are more prominent. Similarly, you can match the height of a nearby countertop, half wall, or bar top.

## TRADE SECRET

To measure the exact distance between two end walls, use pinch sticks—two sticks that slide past each other until the ends hit the walls. Cut the ends at an angle so that they hit the walls at the long point. A pair of hatch marks across the two sticks marks the length. This method is especially helpful when you have to work alone.

A fixed shelf length accents horizontal lines and ties two distinct areas together.

## Shelf dimensions

Once you've identified the general design features, it's time to consider the specific layout of shelves inside each case. Here you must account for what the shelves will hold, as well as for the strength of the materials you intend to use.

**Case width.** The shelf span determines the case width. Pay careful attention to the structure so that the shelves won't sag. The sag of a shelf is measured as deflection, or the maximum distance at which the middle of a shelf bends under the weight of books or other items. The longer the shelf, the more it will deflect; the shorter the shelf, the stiffer it will be.

I aim for a deflection of no more than ¼ in. across the length of the shelf. Therefore, the maxi-

### Shelf Deflection

*The material and length of a shelf affect how much the shelf will sag—or deflect—when it is weighed down by books or other heavy objects. Deflections of more than ¼ in. are noticeable. To reduce deflection, add hardwood banding, use stiffer shelving material, or decrease the shelf span.*

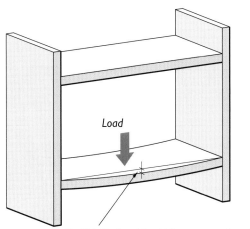

*Load*

*Shelf deflection (should be no more than ¼ in.)*

A variety of short and long shelves provides custom compartments and visual interest.

mum length that a shelf can be depends on the strength of the materials you choose. For example, if you use MDF without banding, the maximum length for a shelf is about 20 in. If you use 1× pine or plywood (also without banding), the maximum length for a shelf is about 34 in. Solid 1× hardwood can span up to 36 in. (See the chart at right for free-span distance between vertical shelf supports.)

Regardless of the material you use, it's best to band the front edges with a 1×2 (nominal) solid-wood apron (preferably hardwood). In essence, the apron acts like a beam, doubling the load-carrying capacity of the shelf. An apron also gives the shelves some visual weight, too, which I think looks better.

**Case depth.** A standard bookshelf measures 12 in. deep. This size holds most books, which range in width from 4 in. for standard paperbacks to 11 in. for binders and reference books. Large art books are often wider. Since the width of the material you use is a limiting factor, it's more economical to let over-size books hang past the edge slightly.

## Droop-Free Shelf Lengths

| Shelving Material | Without Banding | With 1×2 Solid-Wood Banding |
|---|---|---|
| MDF | 32 in. (max) | 40 in. |
| Softwood boards or veneer-core plywood | 34 in. | 48 in. |
| Hardwood boards | 36 in. | 60 in. |

*Note:* Assumes that shelf stock is ¾ in. thick and that the shelves will carry heavy reference books. The maximum deflection is ³⁄₁₆ in. across the length of the shelf.

**Case height.** I usually recommend that vertical cases extend all the way to the ceiling and tie into it with a cove or crown molding. Otherwise, you'll just build a dust trap on top of the case. In actuality however, floor-to-ceiling bookcases aren't built to the exact height available. You must build the cases at least 2 in. shorter than the actual space so that the unit can be tipped into place and lifted up over the hanging rail without running into the ceiling (see Using a hanging rail on p. 141). The empty space will be covered by the fascia and ceiling trim.

**When nailing hardwood banding to the front of shelves, be sure to drive the nail straight, otherwise it will blow out on the face of the shelf.**

### Shelf Banding Options

#### Solid Wood
A nominal 1×2 hardwood apron adds visual weight to each shelf; when glued and nailed to the edge, it dramatically increases the shelf's load-bearing capacity.

#### Wood Moldings
This is often the least expensive option. Off-the-shelf moldings, such as ¾-in.-wide screen bead and 1-in.-wide lattice, can be applied with little effort—no planing or heavy sanding are required. Wider moldings lend visual weight to the shelves, but because these softwood moldings are thin, they don't significantly increase the strength of the shelves.

#### Plastic Strips
T-moldings and adhesive-backed strips can be installed quickly. I prefer T-moldings, which are inserted into a continuous slot in the shelf edge. Adhesive-backed strips can lose their adhesiveness and require additional gluing, which can be aggravating as well as time consuming.

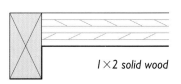

1×2 solid wood

Lattice

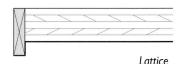

Iron-on (wood veneer or polyester)

T-molding

*When "mapping" a room, always start by checking the floor for level. Remember to check across the entire length, not just across the length of your level.*

## IN DETAIL

Although particleboard may seem like an economical choice for shelving, it's one of the flimsiest materials available. However, you can increase its strength in a number of ways:

- Gluing a ⅛-in.-thick edge band to a ¾-in.-thick particleboard shelf increases the shelf's load-bearing capacity by about 15%.
- Adding 5/100-in.-thick plastic laminate to one edge and both faces increases a particleboard shelf's load-bearing capacity by about 200%.
- Gluing a 1×2 solid-wood apron to the front edge increases its load-bearing capacity by 300% to 400%.

These strategies will work for any material, allowing you to increase the shelf length without creating a sag.

## IN DETAIL

Most plywood sheets are 48 in. wide, producing a maximum of four full-length 11⅞-in.-wide rippings. The extra ½ in. (across the entire panel) is eaten up by the saw, which makes four cuts with a ⅛-in.-wide blade. This is a reasonable width to use for the case sides.

You can build horizontal shelving to any height, depending on your needs. The top might be a useful and needed counter area. Island bookcases can be positioned to screen or otherwise divide a room, while still letting natural light into the partitioned space. In those instances, look at the heights of windows and door headers, sills, sashes, countertops, and other prominent features in the room to determine whether they can be matched. The correspondence between casework and house features can pull together the entire design of a room, lending a degree of intentionality to your work that will be noticeable and visually pleasing.

**Height between shelves.** Adjustable shelving provides a lot of flexibility when it comes to selecting the heights of individual shelves. I almost always design shelving so that most of the inside cases are supported by some kind of adjustable system, such as pins or tracks.

Regardless of whether the shelves will be fixed or adjustable, it's important to measure the books and other items you intend to place on them. Then sketch out what might work. Books range in height from 6½ in. for short paperbacks to more than 12 in for reference guides. Most books are between 8 and 12 in. tall. Add at least 1 in. of extra space so that you have room to tip them out.

## Fitting the space

To fit a bookcase in a room, you also need to consider the outside dimensions of the unit and figure out how much usable space there is in which to work. This sounds simple enough, but there's a catch: You can't simply measure the floor space and the wall area and then determine the overall dimensions. Rather, you must consider all three dimensions in relation to each other.

Here's an example to illustrate what I mean. One of the first large shelf units I built was for a rambling, Victorian house. The room had 14-ft.

## Efficient Plywood Cutting

For a standard bookshelf unit, I try to make the finished shelves as close to 12 in. wide as possible in order to fit large books. Using nominal 1×2 hardwood shelf aprons and a face frame, my rip list for panel stock (MDF or plywood) looks like this:

| Material | Width of Panel Rip | Width with 1× Banding |
|---|---|---|
| Shelf stock | 11⅛ in. | 11⅞ in. |
| Case stock | | |
| End panel | 12 in. | 12¾ in. |
| Scribe panel | 13 in. | 13¾ in. |

ceilings and the customer wanted bookshelves along an entire wall. No problem (I thought). I measured the floor and wall areas and marched off to build the cases. But I didn't check to see how plumb the walls were, how level the floor was, and how square the corners were. The bookshelves didn't fit . . . at all. Not only was the floor 1 in. out of level over 12 ft., but the walls were nearly 2 in. out of plumb over 14 ft. Because I had built my nice square cases to fit the exact floor space and the exact wall area, there was no way I could ever squeeze them into the room.

To avoid this problem, I have learned to do two things when designing floor-to-ceiling built-ins: 1) Map all three dimensions of the room and 2) based on the map, build the cases small and scribe the face frame to fit.

**A 3-d map.** To "map" a room in three dimensions, start by checking the floor for level. If the floor is out of level from the front to the back of the shelving, you will be able to adjust for this when you set the kick base (see p. 138). If the floor is out of level along the length of the shelving, you must obtain an accurate assessment across

# MAPPING THE AREA

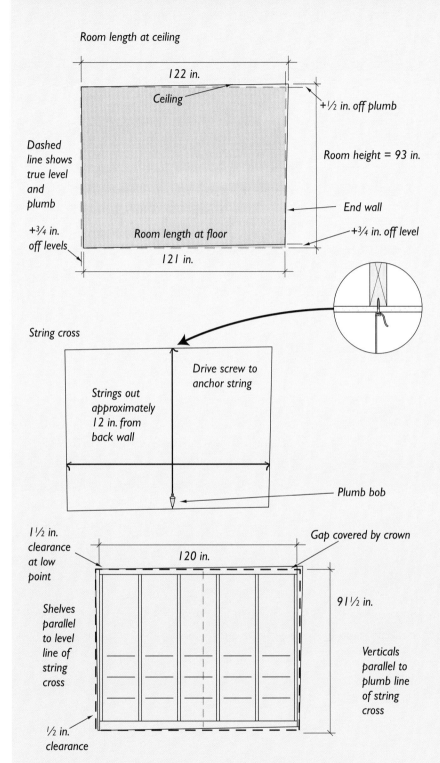

Room length at ceiling

122 in.

Ceiling

+1/2 in. off plumb

Room height = 93 in.

Dashed line shows true level and plumb

End wall

+3/4 in. off levels

Room length at floor

+3/4 in. off level

121 in.

String cross

Drive screw to anchor string

Strings out approximately 12 in. from back wall

Plumb bob

11/2 in. clearance at low point

Gap covered by crown

120 in.

Shelves parallel to level line of string cross

911/2 in.

Verticals parallel to plumb line of string cross

1/2 in. clearance

In old houses with sagging floors, drooping ceilings, and leaning walls, it's important to take your time to "map" the area for a built-in.

1. On a piece of paper, start with an ideal scenario (the dashed line) showing plumb end walls, a level floor and ceiling, and square corners. As you check the floor and ceiling for level and plumb, measure the discrepancies and note them on the map.

2. For large areas, it's sometimes helpful to set up a "string cross" secured with pushpins. This establishes a level and plumb reference from which to pull accurate measurements. Note that the cross does not necessarily need to be located in the center, but it should be positioned off the back wall just inside the face of the built-in.

3. Once you have an accurate picture of the walls, floor, and ceiling, outline the built-in. Make sure the outside line is at least 1/2 in. away from the wall. Note that the base of the built-in is 31/2 in. up from the highest point on the floor. This leaves room for a kick space built on a level 2×4 platform. The shaded area represents the space that will be filled in with a scribe strip (on each end wall) and a fascia (at the head).

### IN DETAIL

The very widest solid stock (1×12) measures 11¼ in. wide. If a nominal 1×2 apron is added to the front edge, you can obtain shelves measuring a full 12 in. wide. Without the apron, however, the shelves will be narrower than those of standard bookcases. And if this stock is used for the sides of the case, the shelves will be narrower still.

### TRADE SECRET

Most of the time, I use MDF panels for paint-grade bookshelves. They come in 49-in. by 97-in. sheets. The extra inch in width allows for four full-length 12-in.-wide pieces, with room to spare for the kerf (the material eaten up by the blade). Particleboard also comes in this larger sheet dimension, but it doesn't machine as cleanly as MDF.

the entire length, not just along the length of your level. Hold a 4-ft. level on a reasonably straight board. I find 1×2 strapping is the easiest to maneuver indoors, but any board that's long enough to span the room will do.

Also check the ceiling for level. If it is out of level as well, determine whether the ceiling "follows" the floor or whether it has its own sag. Similarly, check the walls for plumb and flatness. Assess the wall along the back of shelving and on any sides into which the shelf will butt. If the shelves butt two walls at each end, try using pinch sticks to get an exact and easy measurement.

If all the walls and ceilings are wildly out of level and plumb, it's easy to get confused. To keep track of the directions in which the walls lean and the floors slope, I draw a picture of the entire area. If the area is especially large, or if I need super-accurate numbers, I set up a "string cross."

Using pushpins in the floor and ceiling, stretch a plumb string line. Similarly, using pushpins in the end walls, stretch a level line between the walls. This provides a plumb and level reference from which to pull accurate measurements. The string cross is helpful for identifying "bellies" (or sags) in the floor and ceiling.

Once you have an accurate picture of the space, draw in the level and plumb lines that rep-

resent the edges of the built-in units. Give yourself plenty of leeway here. The unit should be at least ½ in. away from the walls and ceiling at all places. However, this means that there will be a much bigger gap between the walls and the ceiling in other places if the walls are out of plumb and the floor and ceiling are out of level. Not to worry. This gap will be filled in with a scribed face frame.

## Building Cases

Once you have accurate numbers for the outside case dimensions, you're ready to build the unit. The bookshelf unit that I describe here has two main parts—the case, or plywood box, and the shelves that fit inside it. The face frame, kick base, and fascia are done later, when you install the bookcase. Let's start with the case.

### Cutting panels

Begin by ripping the panels lengthwise for the cases and shelves. Plan to cut all of these parts at once, using the same saw settings so that the pieces are consistent. In many ways, consistency is more important than exact, whole number dimensions.

When using veneer-core plywood, pay attention to the orientation of the grain. The grain

**Case Back Details**
*The best way to join the back to the sides of a bookcase is with a rabbet, which holds the back securely in place. However, when the side panel is exposed, cut a scribe edge, as shown, and then glue and nail the back securely to the step.*

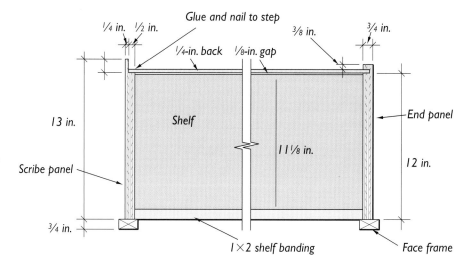

should run the length of the shelf to make use of the plywood's strongest dimension. MDF is equally strong in both directions, so grain orientation is not an issue.

**Side panels.** The side panels are the two vertical sides of the case. I cut two different types, depending on whether a side butts a wall or is exposed.

Sides that butt the walls are not visible, except inside the case. They are also shorter than exposed sides, because they don't need to cover the kick base. Sides that are exposed must be cut longer so that they cover the end of the 2×4 kick base, but they also need to be scribed to the back wall. I call these scribe panels.

I rip scribe panels 1 in. wider than other end panels (typically 13 in. wide instead of 12 in. wide). Then I cut a ⅜-in. by 1-in. step in the panel. I typically cut this step in two passes on the table saw. For the first pass, raise the table saw blade 1 in. above the table and run the panel through on edge (this task usually requires two people). For the second pass, lower the blade to just ⅜ in. above the table and run the panel through flat. If I'm working alone, I often use a router with a straight-cut router bit to cut this step. However, this creates a lot of dust, especially with MDF. A third (and probably easier) method is to use a dado blade on the table saw if you have one.

The step on a scribe panel provides a place on which to nail the cabinet back (typically ¼-in.-

### + SAFETY FIRST

When cutting full, heavy sheets of panel stock with a table saw, get help. Another set of hands to help you support the sheet is better than almost any table support system, unless you have a large shop and have room for an extra-large table.

## Bookshelf Building Checklist

### Panel Cutting

1. Rip panels and shelving stock from full 4×8 sheets.

2. Cut a step along the back edge of the scribe panels.

3. Crosscut the panel stock to length (the sides, tops, and bottoms of the cases).

4. Crosscut the shelves to length (¼ in. less than the inside case width).

### Panel Prep

5. Prep the side panels for biscuits.

6. Prep the side panels for adjustable shelves (drill holes for shelf pins or rout channels for metal standards).

7. Cut a notch for the kick space in the exposed side panels.

### Case Assembly

8. Glue the biscuits and assemble the case (nail the exposed sides and screw the sides that abut the walls).

9. Glue and nail the back.

10. Screw on half of the museum rail to the cabinet's back.

thick luaun plywood), leaving a ⅜-in.-thick "fin" that covers the gap created by the hanging rail (see pp. 141–142 for more information on hanging rails). In old houses, the back wall may be out of plumb or the surface out of flat. If that is the case, rip the scribe panel wider, so that the fin can be scribed to match the tilt and wave of the back wall.

**Chopping panels to length.** Once the panels have been ripped, I chop them to length, usually with a large sliding miter saw. Most sliding chopsaws can cut materials up to 12 in. wide, so they can easily handle the shelf stock, side panels, and top and bottom panels of cases. However, they don't quite cut all the way across

## PRO TIP

*Rare is the old house that doesn't have sagging floors and ceilings. In my experience with older homes, floors and ceilings often sink uniformly toward the stairs.*

## IN DETAIL

There are essentially two systems for making adjustable shelves in built-ins: standards and pins.

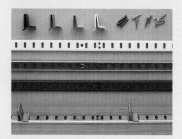

- **Metal standards** are nailed or screwed in pairs on both sides of each shelving unit. They are much more conspicuous than pins, but they are also stronger and generally much faster to install. They look best when let in to dadoes.

- **Shelf pins** come in a wide variety of styles. I prefer the spoon type, which are the least conspicuous. All types fit into a series of holes drilled into the sides of the bookcase. Drilling the holes can be time consuming.

wider scribe panels. For those, I usually cut as far as I can. Then, carefully supporting the cut, I slide the panel to the edge of the chopsaw table and finish the cut with a handsaw or cordless panel saw. Because of the difference in blade thickness between the chopsaw and whatever else you use to finish the cut, there is often a little nubbin of material left. I use a utility knife or a sharp 1-in. chisel to very carefully pare off this bit of material, leaving a single, clean surface on the cut end.

Of course, you don't need a sliding chopsaw to make these crosscuts. You can also make them with a radial-arm or circular saw. I am not a big fan of radial-arm saws because I feel that they are too dangerous. But they have ample crosscut capacity for this type of work and cut with very little tearout, so many carpenters would argue that, in this case, they are the only practical choice.

### Two Case Joints

#### Tongue-and-Groove Case Joint
*A table saw with a dado blade can cut this sturdy joint in a few passes.*

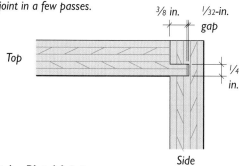

#### Biscuit (or Plate) Joint
*The modern biscuit joint is strong and quick to set up. Wood biscuits fit into slots cut by a plate joiner and swell on contact with glue, creating a secure and accurate joint.*

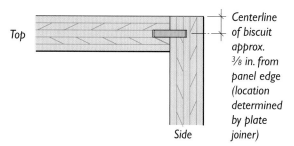

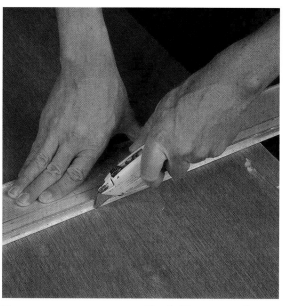

**To prevent tearout, use a utility knife to score the cut deeply enough to break the grain at the surface.**

Using a circular saw takes the most time, because you need to set up for each cut. Make sure that the panels are supported on both sides of the cut (I use a pair of 2×4s spanning a pair of sawhorses to support the panels along their full length). On veneered panels, you will probably have to score the cut with a utility knife to prevent tearout. For chopping shelves and the top and bottom pieces of the case, you will probably have to score both sides of the cut, since the end of one cut is the beginning of another on the next panel. This requires patience and precision. Using the thinnest saw blade with the most teeth helps

### + SAFETY FIRST

Radial-arm saws can be very dangerous because they are pulled across the cut (most sliding miter saws are pushed across the cut). If the blade jams, it can easily climb out of the cut and head straight at the user. Make sure your digits and limbs are safely out of the way at all times. The carpenters I know who are missing fingers lost them to radial-arm saws during a momentary lapse of attention.

reduce tearout. For a 7¼-in. saw, for example, use an ultra-thin, 36-tooth blade. I find a 24-tooth blade on a cordless panel saw works well.

If the floor slopes out of level from the front to the back of the case, be sure to leave the exposed side panels long enough. I allow 4 in. to 4½ in. of extra length to cover the kick base. Since the kick base is built from 2×4s, which are only 3½ in. wide, this method provides an extra ½ in. to 1 in. for cutting the exposed panel to match the floor.

## Prepping panels

After the panel stock has been ripped and chopped to size, I prep the panels by cutting slots for biscuits at the joints and drilling holes for shelf pins.

**Biscuit joints.** There are two ways to secure fixed shelves and join the sides of a case to the top and bottom pieces. I used to cut tongue-and-groove case joints, but now I use a plate joiner (see the sidebar on p.135).

Start by marking the location of the biscuits on the inside face of each side panel. Measure the location of each intersecting panel (top, bottom, and any fixed shelves). With a framing square, draw a line on both sides of the intersecting panel. I use these intersection lines for aligning the faceplate jig when cutting slots in the face of the side panels.

**Shelf-pin holes.** Once all the biscuits are marked and cut, lay out the shelf-pin holes. I use a

### + SAFETY FIRST

It may sound redundant to say "wear safety glasses," but this advice bears repeating. Veneered plywood in particular splinters easily and throws back a lot of chips. Don't be caught blind without eye protection.

## Sample Case Cut List

1. Side panels at wall

   - Rip width to 12 in.

   - Cut to the height of the bookshelf, less the kick height.

2. Exposed (scribe) side panels:

   - Rip width to 13 in.

   - Cut to the height of the bookcase from the floor (including the height of the kick base)

3. Top and bottom:

   - Rip width to 12 in.

4. Fixed shelves:

   - Rip width to 12 in.

   - Cut the length to the inside width of the case.

5. Adjustable shelves:

   - Rip width to 11⅛ in.

   - Cut the length to the inside width of the case, less ¼ in.

6. Back:

   - Rip the width to the outside width of the case, less ⅜ in. (for each scribe panel).

   - Cut to the height of the bookshelf, less the kick height.

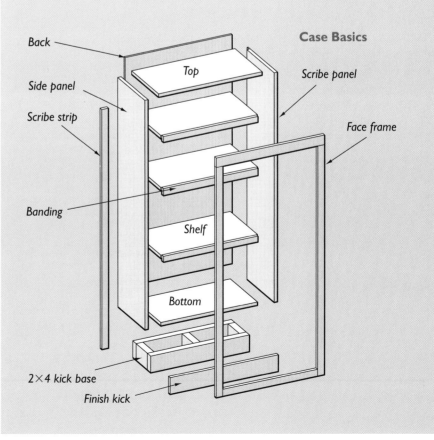

**Case Basics**

Back · Top · Scribe panel · Side panel · Scribe strip · Face frame · Banding · Shelf · Bottom · 2×4 kick base · Finish kick

**PRO TIP**

*When cutting multiple parts, consistency is more important than exact measurements.*

## WHAT CAN GO WRONG

Wood grain that splinters along the edge of a cut is known as "tearout." True to its name, tearout is the result of the saw's teeth coming out of the cut and tearing the grain as it leaves the wood. Tearout is predictable and likely to occur in these situations:

- **Orientation of the grain:** Tearout is always worse when crosscutting and minimal when ripping.
- **Solid wood:** Open-grained woods, such as oak and ash, are more subject to tearout than close-grained woods, such as birch and maple.
- **Veneer:** All veneers are prone to tearout when they are crosscut. The thinner the veneer, the more likely it is to splinter along the cut.

Use a sharp pencil to mark a crisp carat at each shelf-pin location.

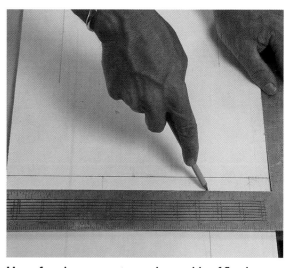

Use a framing square to mark one side of fixed cross panels.

small tri-square and mark a line 2 in. in from each long edge of each side panel. Mark the inside face of the panels; on scribe panels, be sure to measure from the step.

I usually space holes 2 in. apart. Starting from the top edge of the side panel, I begin my layout 8 in. down and end about 12 in. from the bottom. If I have fixed shelves, I stop and start the series about 6 in. on each side of the intersection lines.

Drill the holes with a brad-point bit to match the diameter of the pins (usually ¼ in. to ³⁄₁₆ in.). Use a drill stop on the bit so that you can repeat-

edly drill to a consistent ³⁄₈-in. depth. Laying out and cutting shelf-pins holes this way can be tedious work, and there are several jigs you can make or buy to speed the process along. But I find that if I'm building one case at a time, it's just as easy to lay out and drill by hand as it is to make and store a template.

**Kick space notch.** The final step in preparing the panels is to cut a notch in the exposed side panels. I cut the notch with a jigsaw, making it 3 in. deep (that is, 3 in. from the face of the case) and 4 in. to 4½ in. high.

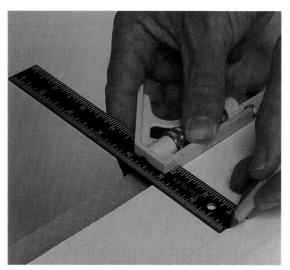

A tri-square provides a consistent measure from the panel edge for each shelf pinhole.

A **Speed Rip Square** provides a ready-made guide for cutting a crisp knockout for the kick space on the finished end panel. Make sure you leave the bottom long, so that it can be cut to the slope of the floor.

## Assembling the cases

After the panels have been cut to size and prepped for biscuits and shelf pins, you are ready to assemble the cases. Once the glue begins to cure and the biscuits start to swell (this takes less than 10 minutes), there's no going back. So before you begin, make sure that you have everything you need on hand. This includes:

- Wood glue.
- A wet sponge and bucket of water for wiping up excess glue.
- A sharp chisel for scraping glue out of corners.
- A cordless screw gun and plenty of 1⅝-in.-long wood screws (square drive preferred) for securing joints on panels that abut walls.
- A nailer and 1½-in.-long nails for securing joints on exposed sides.

Start by transferring the centerlines of all the fixed shelves to the outside face of the panel. This provides an exact location for screwing and nailing. Then you're ready to apply the glue.

**Gluing the biscuits.** Squirt a liberal amount of glue into each slot on one joint. Insert a biscuit and smear the squeeze-out along the joint. It's easy to apply too much glue (but important to apply

**Do not use too much glue when setting biscuits.**

## Plate-Joining Techniques

The fence and blade housing of a plate joiner have centerlines marked prominently in red, so layout is simply a matter of aligning this centerline with the centerline of the slot you want to cut. The easiest way to lay out biscuits is to align the panels as they will be joined, and then mark the biscuit that will hold them together with a single slash.

While most plate joiners have elaborate adjustable fences, I rarely use one. Instead, I work right from my work surface. Without the fence, most plate joiners are set up to cut a slot about ⅜ in. up from the bottom of the tool—approximately in the middle of ¾-in.-thick material. Keep track of which is the bottom and which is the top of each panel, because not all plate joiners cut a slot dead center.

To cut a biscuit slot in the edge of a panel, hold the plate joiner flat on the work surface and push the tool into the edge of the panel. Keep all of the intersecting panels face up (as they will be installed), so that the slots will be in a consistent location.

To cut a slot in the face of a panel, I use a spring clamp to secure an angle jig to the side of the panel. The bottom of the plate joiner rides on the vertical leg of the jig. I always clamp the jig to the bottom of the intersecting panel so that the face slots align perfectly with the edge slots.

## IN DETAIL

Tearout usually occurs on only one face—the face where the blade exits. Therefore, the best way to avoid tearout is to support the grain on the "upcut" side of the cut, which depends on the rotation of the blade.

If the grain cannot be effectively supported, score along the cut with a sharp utility knife. When using a circular saw, you may need to both score the cut on one side of the blade and support the grain on the other side of the blade with a shoot board.

*Circular saw*

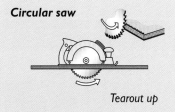

*Tearout up*

*Table saw*

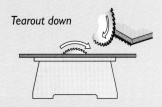

*Tearout down*

If the side of the shelf unit will not be seen, it can be secured with screws, which hold better in the panel end grain.

On an exposed end panel, however, use finish nails for a cleaner appearance.

Begin by gluing the cross panels—the top, bottom, and fixed shelves—to one side panel, working with all of the pieces on edge. Once glue is applied to the biscuits, they will swell within about 10 minutes, so make sure that you're organized and able to work quickly.

enough), which is why it's imperative to have a sponge and bucket of water on hand. At this stage, don't glue more than one joint at a time.

When the biscuits are firmly pressed into their slots, turn the panel on edge, align it with the end of a fixed shelf, and push the two meeting pieces together. Assembling a case with biscuits is sort of like working with Legos—the pieces almost snap together. But you must work fast and remain focused. As the glue dries, the biscuits swell, which helps clamp the joint fast. Before that happens, make sure that you've pulled the two meeting pieces together and screwed or nailed them tightly together.

After the first joint is glued and secured, all of the remaining joints must be glued with the panels on edge. I glue all of the joints on one side panel first, and then turn the side panel flat on the bench so that the attached pieces are sticking up in the air. Then I glue all of the remaining joints at once, lift the second side panel into place, and align each joint until the biscuits find

After the cross panels are affixed to one side, flop the side panel onto the bench so that the cross panels stand upright, and then drop the second side panel on top of them.

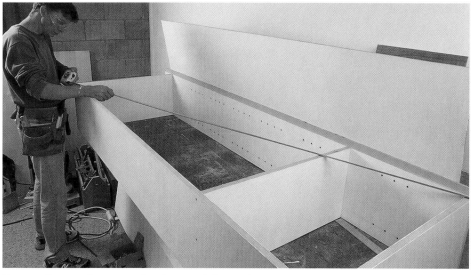

Square the case by measuring for equal diagonals.

Before nailing the back, apply a light bead of wood glue to the case edges.

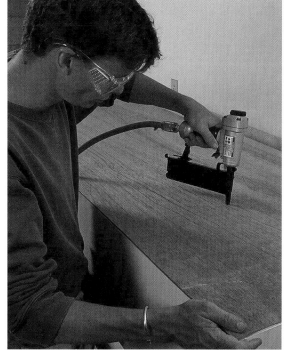

Once the case is square, the back can be nailed in place.

their slots and fall into place. This can be nerve-racking to do all at once, so it helps to dry-fit the second panel before glue-up so that you can be confident it will fit when the glue is on.

**Installing the back.** At this point, all sides of the case are glued and secured. Lay the case on its front so that the back edges are pointing up. Begin by squaring the unit: Measure the diagonal distance (from corner to corner) across the case, being sure to measure inside the step if you have one or more scribe panels. Measure both diagonals, which should be equal. If they aren't, push

the case around until they are. When the case is square, apply glue to the edges, carefully lift the back into place, and nail it off.

A continuous bead of glue is especially important along all of the back edges, so that books and other objects can be pushed hard to the back without opening an unsightly gap. I've found that, in addition to glue, headed ring-shank nails work best for securing the back to the case.

## PRO TIP

*A hanging rail may seem like an unnecessary extra step, but it speeds installation and keeps the case from tipping.*

## IN DETAIL

Stud sensors can detect wood studs in conventional interior walls by sending out a signal and then reading the echo that bounces back. The interval of the echo varies depending on the density of the material. To use a stud sensor, first calibrate the device over an empty wall cavity (an area of the wall that doesn't have framing beneath the drywall). When a calibrated sensor is moved over framing, it lights up, indicating that the signal has bounced back immediately from a denser material. Stud sensors don't work when foil-faced insulation materials (foam panels or batts) are used, since the foil reflects the signal evenly across the wall.

## Fastening Panel Stock

Biscuits provide the strongest connection between two panel pieces. Nails or screws supplement this fastening, but they are used mostly for holding the joint tight until the glue around the biscuit dries.

- Use 1½-in. nails to clamp the joint on exposed faces. I typically use a pneumatic nailer, which drives the nail in one clean stroke. Driving nails by hand can knock the joint apart as you strike repeatedly with the hammer. If you do use a hammer, predrill the holes to minimize movement.

- A better alternative to hand nailing exposed faces is to use finish head screws. These require more work to fill the holes over the screw heads. Drive the head about ⅛ in. below the surface. It's easy to strip out the small head, so predrill the holes and make sure you use the proper drive bit (a #1 Phillips head or a small square drive).

- Use screws on faces that won't be seen (for example, when they face a wall or another case). Screws grab better in the edges of all panel materials, drawing the joint tighter. I prefer using square-drive screws, which don't cam out as easily as Phillips-head drives.

- Use 4d ring-shank nails (hand-driven) or narrow-crown staples (pneumatic driven) to secure case backs.

# Installing Bookshelves

The method I describe here for installing bookcases has two essential components: a base and a hanging rail. The base is necessary for supporting a heavy load of books, and a hanging rail is important to prevent a full bookcase from being pulled over onto someone—a very dangerous possibility. But please note: I make my base and hanging rail in a special way, which I find works well. However, you should feel free to adapt these methods as you see fit. What is important is that you provide a sturdy base and a secure wall attachment for built-in bookcases.

## Kick base

I typically install my bookcases on a kick base. That is, the bottom of the front face is recessed 3 in., much the same way in which kitchen cabinets are built. This is by no means mandatory. In fact, many customers prefer the base to be wrapped with baseboard, rather than having a recessed base as I describe here. You can easily adapt this method to your personal preferences.

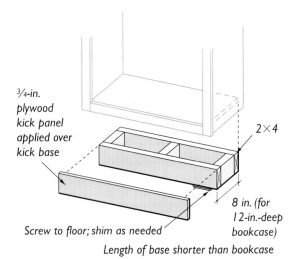

¾-in. plywood kick panel applied over kick base

2×4

8 in. (for 12-in.-deep bookcase)

Screw to floor; shim as needed

Length of base shorter than bookcase

**Leveling Base**
*Build a solid leveling base out of 2×4 stock. The base should be smaller than the footprint of your bookcase. A ¾-in. plywood kick panel can be applied over the front after the cases have been installed.*

You can even build the recessed base and then wrap it with baseboard if you don't like the way it looks. The important thing is that you build a level base for supporting a heavy load of books.

**2×4 ladder.** Begin by building a 2×4 ladder. For a standard 12-in.-deep bookcase, I build this base about 8 in. wide and several inches short of the total bookcase length. It's much easier to build the structural base small so that it doesn't bump into old shoe molding and baseboard that may remain behind the built-in. If you have several units sitting side by side along a wall, build one "ladder" base for the entire run. This vastly simplifies the installation.

Start by snapping a chalkline on the floor where the front edge of the base will be. If the base is recessed, this line will be 3 in. from the front face of the unit. Refer to your 3-d map (see Mapping the Area on p. 129) to make sure that the case can be set plumb without running into the back wall higher up. If you're not sure, test-fit the location of the base by lifting a case onto the ladder base and checking for plumb.

When you're certain of the location of the front edge, level the ladder base and secure it to the floor with 3-in.-long wood screws. Be sure to check for level in both directions—along the length and from front to back. Shim the base with solid shims at each point where it is screwed to

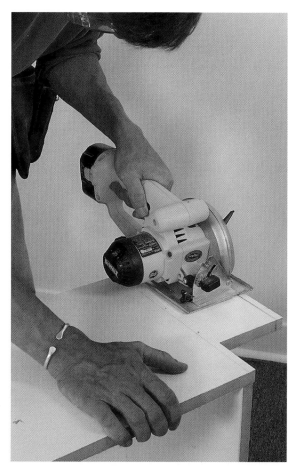

The bottom of a finished end panel should be cut to its final length in place. Although you can use a circular saw, a jigsaw tends to kick up less dust.

## The Art of Scribing

Scribing is a method of transferring the contour of a wavy wall, ceiling, or floor surface onto woodwork. Once marked, the woodwork can be installed without any gaps showing. This method takes a lot of practice to do well, and your first few attempts may still require a little caulk (affectionately called "tube scribing") to dress out the joint.

To transfer the contour of a wall, use a drawing compass. Most carpenters I know rely on an inexpensive school compass, though many soon graduate to fancy wing dividers. The most important feature of any scribe tool is that the two legs can be securely tightened so that the distance between them doesn't change.

Position the panel so it is plumb (or however you want it in its final installation), and press it as tightly as possible to the surface it will meet. The panel may meet the surface at just one point or it may meet in several places. Choose the widest gap and set the dividers for that distance. Then, holding this distance consistent, run the scribing tool along the surface, drawing a line onto the panel. This line will mirror the surface.

When scribing, consistency is everything. Hold the scribing tool at a constant position. I usually hold it parallel to a meeting surface (for example, level with the floor or ceiling when scribing a wall panel) and lightly rest my fingers on the wall to hold the scribe at a consistent angle as I draw (see the photos on p. 140).

When you have a scribed line, cut the panel with a jigsaw. You may have to fine-tune the cut several times, holding the work against the surface repeatedly, resetting the scribing tool to the widest gap, and rescribing the line until you have closed every gap. A low-angle block plane works wonders for this fine-tuning work.

## IN DETAIL

Stain-grade hardwoods add an extra touch of interest and elegance to standard paint-grade cases. You can dress up built-ins even more by using contoured face frames. For example, after installing the face frame, use a cove bit in a router to dress up the inside corners.

## IN DETAIL

When choosing a caulk gun, look for these two indispensable features:

1. A built-in spout cutter. Using a built-in spout cutter is safer than whittling off the end with a utility knife and it provides greater control over the size of the hole.

2. A fold-out wire to puncture the seal on a caulk tube. This feature saves you the trouble of searching for a nail that is long enough to reach the length of the spout.

Set scribe to the widest gap.

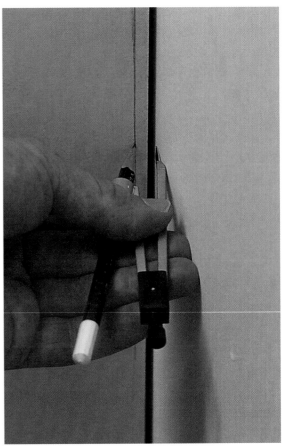

Trace outline of wall on the panel.

A museum rail is ripped from a single board with the table saw set at a 45-degree bevel.

the floor. Use plywood blocks to build up any shim height greater than ½ in.

Once you have the ladder base in place, you will know exactly where the bottom of your bookcase sits. Measure each end from the top of the base to the floor to determine where to make the final cut on the exposed panels.

## Scribing end panels

Before you secure the case to the wall, you may need to scribe-fit the end panels. If there is more than one case per ladder base, start by scribe-fitting an end unit. Place the cabinet on the base and check for plumb. The step that was cut on the scribe panel prevents the unit from sitting in its exact location, so align it parallel to the front of the base, with the scribe fin pushed as close as possible to the wall. If the back wall is out of plumb, you will clearly see the discrepancy. Adjust your scribe tool to the widest gap and trace the wall contour. Remember to hold the scribe parallel to the floor.

It is very important to leave exactly ¾ in. of material (the thickness of the hanging rail) at the level on the wall where the hanging rail will be installed. This can be tricky, but it's critical. If you don't have enough material here, you can end up with a gap worse than the one you started with, and if you have too much, the hanging rail will hold the case higher than it should be.

## Using a hanging rail

There are two big advantages to using a hanging rail. The first is a matter of safety. The rail prevents the bookcase from tipping over—no small detail, especially when there are kids in the house. The second is a matter of convenience for the installer. The rail makes it easy to attach the bookcase to the wall framing without riddling the back with screw heads. Also, because the rail spaces the back of the bookcase at least ¾ in. from the back wall,

it allows you to bridge any surface irregularities that may pull the case out of plumb when you secure it to the back wall.

The rail that I prefer is commonly used by museums and galleries to hang heavy pieces of art. Hence, I typically refer to it as museum rail. It's made by taking a strip of ¾-in.-thick panel stock and ripping it in half lengthwise at a 45-degree angle to create two pieces of rail stock, as shown.

Using 1⅝-in.-long bugle-head wood screws, I attach half of the rail to the back of the bookcase. The 45-degree angle essentially creates a

**One side of the rail is applied to the back of the case (the long point of the bevel faces downward).**

---

### + SAFETY FIRST

A standard throat plate on a table saw typically has a wide gap around the blade. Narrow strips are at risk of falling though the gap or catching on the rim. To avoid a possible kickback, make a zero-clearance throat plate from a piece of plywood or melamine. Remove the old plate and trace it so that the new plate has the exact same perimeter dimensions. Make sure the screw heads holding the old plate are recessed below the top surface of the new plate. To install the new plate, drop the blade all the way down, and then raise it while the saw is running to cut a perfect kerf with no empty space around the blade.

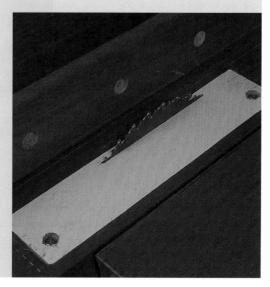

## PRO TIP

*One way to get moisture out of biscuits is to bake them for a few minutes in the oven.*

## IN DETAIL

Plate joiners offer a quick way to join panel stock with minimal setup. Plate joiners are essentially small circular saws designed to cut a semicircular kerf in the edges of wood. The kerf holds a small football-shaped spline (called a "biscuit") made of slightly compressed beech and crosshatched with grooves to hold glue. The biscuit swells in contact with any water-based glue, ensuring a strong bond and a very tight joint. The grain of the biscuit lies at about 30 degrees diagonal to the length, which provides incredible resistance to shearing. Trying to break a biscuit joint usually results in fracturing the surrounding wood, not the biscuit.

## TRADE SECRET

When using a circular saw or router to cut full sheets, support both sides of the cut with 2×4s. I find it easiest to lay down four long 2×4s on the ground.

**The matching side of the rail is screwed to the wall (the long point of the bevel faces upward).**

hook that meets its mate, which is secured to the wall. The height of the rail is not critical. It should be in the upper half—preferably in the upper third—of the case. I try to align the rail with a fixed shelf to provide a good screw base. If that is not possible, glue the rail to the cabinet back.

Next, measure along the back of the case to find the exact height of the meeting rail—the half that attaches to the wall. Be sure to precisely measure this distance on the cabinet. Measure both ends of the case rail, and if it is slightly skewed on the case, orient the wall rail accordingly. When transferring these measurements to the wall, measure from the top of the kick base.

Before screwing the meeting rail to the wall, locate the studs, either by sighting from existing nail locations or by using a stud sensor. Fasten the rail to the wall with 3-in. screws in every stud.

Now you can lift the cabinet onto the base. Push the case until the two sides of the rail bump into each other, and then lift the case up and over the rail. Here is where the case height is most critical—that extra 1 in. to 2 in. provides enough

### Museum Rail

*This type of hanging rail, commonly used by museums and galleries to hang heavy pieces of art, is made from a strip of ¾-in.-thick panel stock ripped in half lengthwise at a 45-degree angle. One length is attached to the back of the bookcase and the other to the wall, creating a kind of hook. Pay attention to the direction of the 45-degree angle when securing each length of rail.*

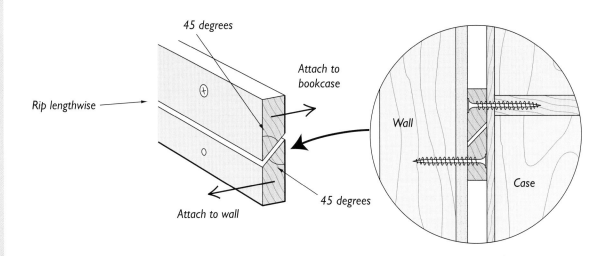

45 degrees

Attach to bookcase

Rip lengthwise

Wall

Case

45 degrees

Attach to wall

**Face Frame Details**

*The ideal width for a face frame's verticals and fixed horizontals is 1½ in. The fascia, which overlaps the verticals across the top, is typically wider (approximately 2 in.). Shelf banding, which is often narrower than the face frame, is typically 1 to 1¼ in. wide.*

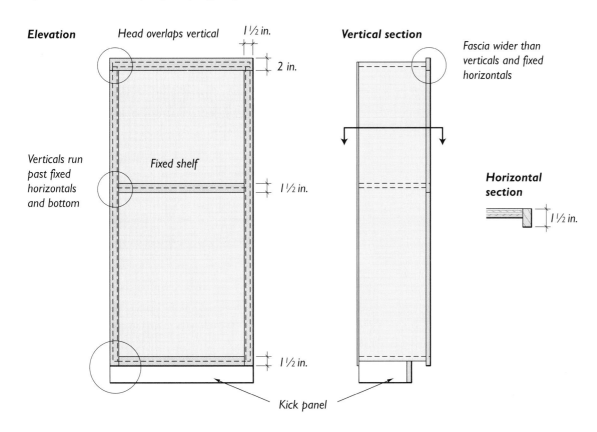

**Elevation**  Head overlaps vertical  1½ in.

2 in.

*Verticals run past fixed horizontals and bottom*

Fixed shelf

1½ in.

1½ in.

Kick panel

**Vertical section**

*Fascia wider than verticals and fixed horizontals*

**Horizontal section**

1½ in.

leeway to lift the case over the top of the 45-degree cut. Once over and in place, however, it's there to stay. Finally, drive screws through the bottom panel into the ladder base to secure the case.

**Finish kick panel.** The finish kick panel can go on anytime after the case unit has been secured. I prefer using a piece of ½-in.-thick MDF, but it isn't really practical to buy a sheet of ½-in.-thick stock when you need only a few pieces. I often use ¾-in. material left over from building the cases.

If the cabinets run from wall to wall, the kick can be cut slightly smaller than the kick space height and scribed to fit. The gap at the top won't show unless you get down on your hands and knees. This gives you the room needed to scribe.

For high-end stain-grade work, I cut a miter on the exposed panel, which meets a miter at each end of the finish kick base to form a clean return.

## Face frame

After the built-ins have been installed, you're ready to dress out the cases with a face frame and ceiling trim. Installing the face frame is finish work at its finest. This is when all the unsightly gaps in an out-of-plumb and out-of level room are covered up and dressed out.

I typically use 1½-in.-wide stock to wrap the edges of the cabinet. However, on the sides that butt into the wall, where the face frame will be scribed to match the contour of the wall, this trim may be slightly wider or narrower.

## PRO TIP

*Make a template from bead board to help locate shelf-pin holes accurately.*

## TRADE SECRET

When drilling holes for shelf pins, it's worth using the very best brad-point bit you can find. Good brad-point bits have a sharp center point, as well as sharpened flute tips to cut a clean circumference around the hole. Even so, it's important to feed the bit at a moderate speed, especially when cutting through the first layer of veneer.

Avoid the temptation to use just any old drill bit for this task. The bit will wander off your mark (causing the pins to become misaligned and the shelves to wobble) and the sides of the hole will tear out, especially in veneered panel stock.

I always install the pieces in this order:

**1.** Start with the fascia, since all of the verticals butt into this piece. If a built-in extends to the ceiling, the fascia wraps around the open ends and may intersect with existing ceiling trim, so you'll need to scribe to match the trim. To wrap the fascia, begin on the exposed side of the bookcase. Cut the side fascia long, trace (or scribe) the ceiling trim profile on the end, and then cut it out with a jigsaw. After checking the fit and marking the length for the front miter, cut the fascia to length and seal the end with adhesive caulk before nailing in place.

**2.** Install the bottom rail, because this piece spans the entire bottom of the case and the verticals "stand" on it.

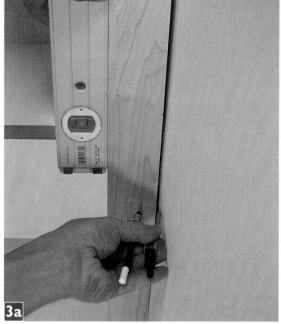

**3.** Install the verticals that must be scribed to fit the side walls. Check the cabinet for plumb, hold the trim piece in place (keeping it plumb, just like the cabinet), and scribe it. Cut the scribed edge with a jigsaw and smoothe the edge with a block plane.

**4.** Install the verticals at the edges of the exposed panels and fill in with the horizontal rails on the fixed shelves.

**5.** I typically install paint-grade trim with adhesive caulk, and while I'm at it, I fill in open corners and any gaps in the case. This saves time when preparing for the final coat of paint, because I am already aware of the blemishes and gaps that I found during the installation process.

# Built-In

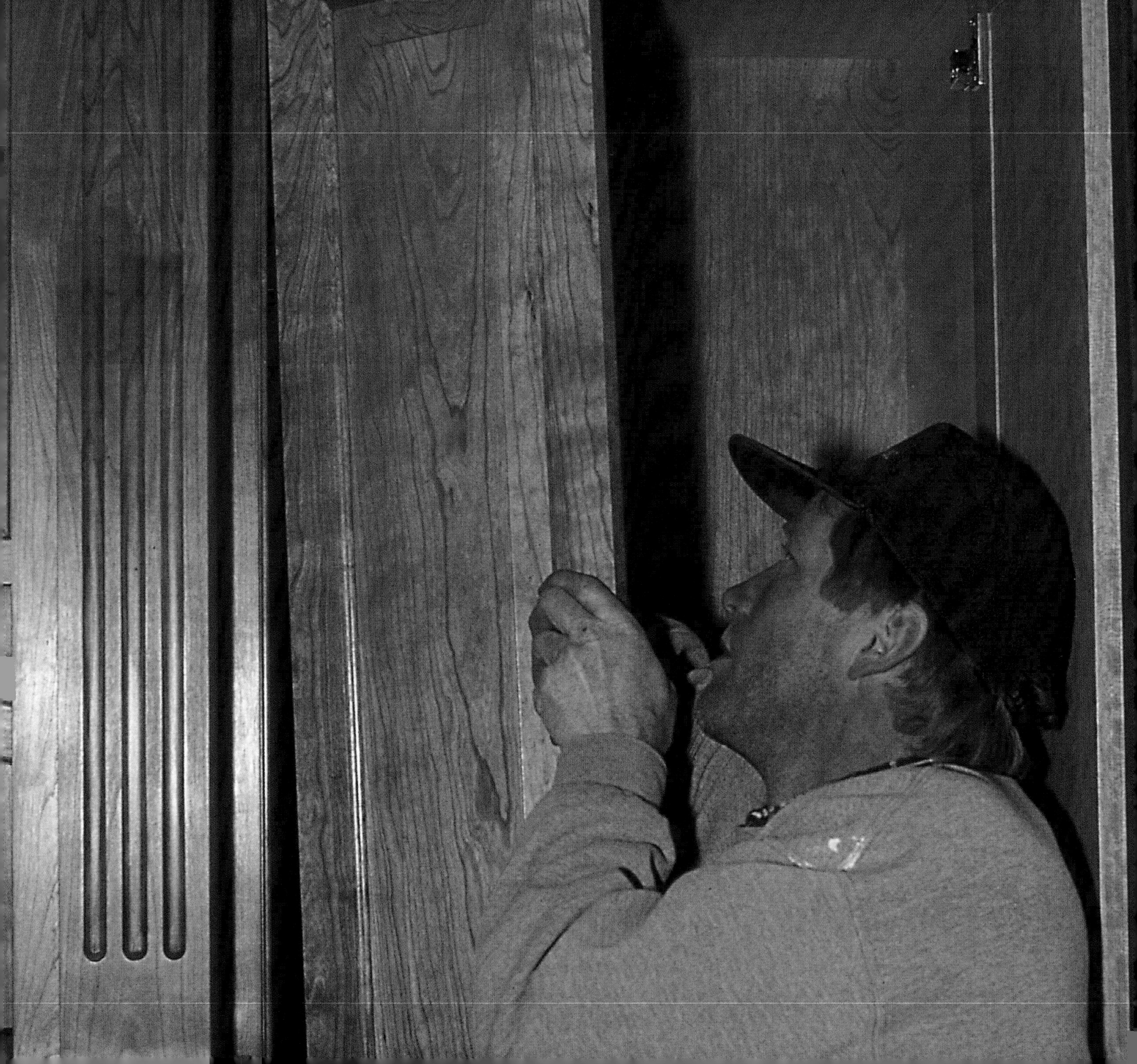

# CHAPTER NINE
# Cabinets

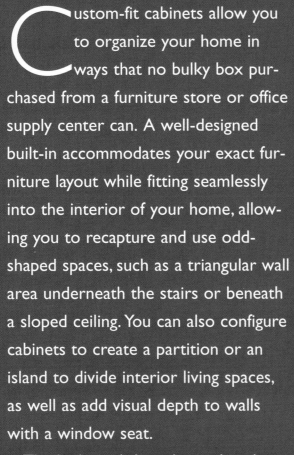

Custom-fit cabinets allow you to organize your home in ways that no bulky box purchased from a furniture store or office supply center can. A well-designed built-in accommodates your exact furniture layout while fitting seamlessly into the interior of your home, allowing you to recapture and use odd-shaped spaces, such as a triangular wall area underneath the stairs or beneath a sloped ceiling. You can also configure cabinets to create a partition or an island to divide interior living spaces, as well as add visual depth to walls with a window seat.

The built-ins I describe in this chapter are an elaboration on the basic box at the heart of the shelving described in chapter 8. By adding doors, drawers, and countertops to this simple plywood case, you can create highly functional storage cabinets, window benches, and built-in workstations.

147

## PRO TIP

*For a wide cabinet, use double doors. As a rule, any cabinet wider than 2 ft. needs double doors both for aesthetics and for relieving the strain on the hinges.*

### IN DETAIL

Pocket-hole joinery is one of the most effective ways to fasten face frames, kick panels, counter splashes, and scribe fins to cabinets. Essentially, the "pocket" is a steep counterbore (drilled either by hand or with a pocket screw cutter) that threads a screw from the face of the ¾-in. (or thicker) board through the edge. This way, the screw won't blow out the finish side. The screw pulls the pieces together and holds fast for an immediate bond, so you often don't need to use clamps.

**Built-in cabinets can fill unused spaces, adding utility as well as decoration.**

# Designing Cabinets

**One of the advantages** of built-in cabinets is that they become an integral part of the house. You can build cabinets to fit the exact interior space, blend finishes with other woodwork, and shape and style the cases to reflect prominent architectural features in the room.

Built-in cabinets can serve myriad needs, including storage for games, art supplies, and household linens or housing for an entertainment system and the attendant CDs and videocassettes. Adding a desktop or a benchtop to a series of cabinets creates an in-home workstation or a window seat. All of these can be built from a combination of modular plywood cases.

The method of cabinet construction that I describe here is often referred to "case-component construction." The built-in cabinets are designed as a series of small, independent components—each one a simple plywood case—stacked or lined up next to each other, filled with shelves or drawers, and overlaid with doors. Typically built as frameless cabinets, the components rely on biscuit joinery and European-style cup hinges and drawer slides. This type of cabinet has no stiles and rails to restrict access to the interior.

## A combination of upper units and base units

Unlike floor-to-ceiling bookshelves, most of the built-in cabinets I install have a lower, or base, unit and a separate upper unit. Often, the upper units are open bookshelves, while doors cover the base units.

**An elegant built-in cabinet visually ties in with the room's panel wainscoting while providing much needed bookshelf and cabinet space.**

Case cabinet construction methods can be adapted to build a window.

The height of the lower cabinets corresponds to the level of the handrail and wainscoting.

Upper shelf units situated above an enclosed base unit provide library shelving as well as a TV nook.

The division between upper units and base units typically falls at a height of 30 in. to 34 in. above the floor, which is a standard desktop or countertop height. The top of the base cabinets can be a solid-wood countertop, providing a strong band of natural wood that visually ties in with wainscoting and handrails. However, the divider between upper units and base units need not be one continuous countertop that separates open shelving from closed cabinets. In fact, variations in open and closed spaces can enhance a wall unit.

## IN DETAIL

For drawers below a counter-top, leave at least ¾ in. of clearance between the top of the drawer box and the top of the cabinet. This allows the drawer to be tipped into and out of the slide without bumping into the stretchers.

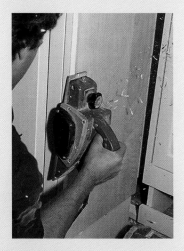

## IN DETAIL

Unlike a jointer, a handheld power plane won't take a bow out of a board, but it will shave off as much as ³⁄₁₆ in. of wood (some bite deeper, but the deeper the bite, the rougher the cut is likely to be). I mostly use a power plane for quickly planing scribed pieces and for cleaning the saw marks off the edges of boards.

## Door layout

In case-cabinet construction, the doors attach to the sides of the cabinet case or to fixed vertical dividers within the case. The door layout looks best when the upper and lower units align verti-cally. For example, notice how each of the shelf bays in the bottom photo on p. 148 mirrors the door layout on the base unit below it.

If the cabinet opening is wide, consider covering it with double doors. I avoid single cabinet doors wider than 2 ft., because they become ungainly to open and too heavy for standard hinges to support. If you're lining a hallway or upstairs balcony area with built-ins (see the bottom left photo on p. 149), leave enough room for the door to swing fully open without running into a wall or balustrade.

### Case Cabinet Construction
*A typical built-in cabinet consists of a base unit below a countertop, with shelf units above.*

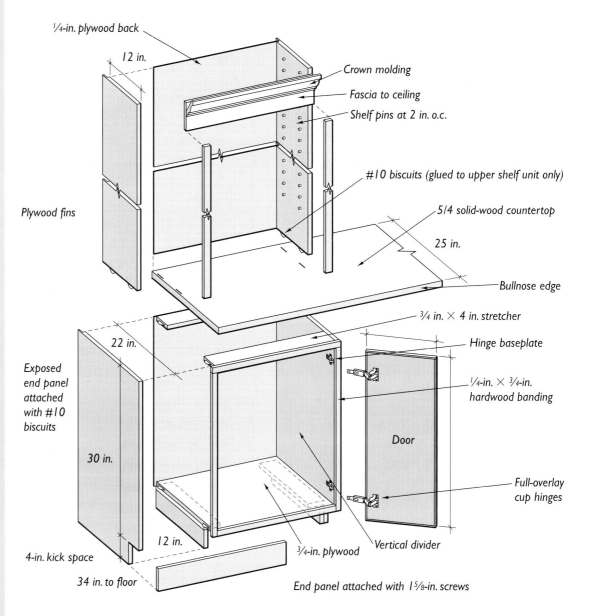

Labels:
- ¼-in. plywood back
- 12 in.
- Crown molding
- Fascia to ceiling
- Shelf pins at 2 in. o.c.
- #10 biscuits (glued to upper shelf unit only)
- 5/4 solid-wood countertop
- 25 in.
- Plywood fins
- Bullnose edge
- ¾ in. × 4 in. stretcher
- Hinge baseplate
- ¼-in. × ¾-in. hardwood banding
- Exposed end panel attached with #10 biscuits
- 22 in.
- 30 in.
- Door
- Full-overlay cup hinges
- 4-in. kick space
- 12 in.
- 34 in. to floor
- ¾-in. plywood
- Vertical divider
- End panel attached with 1⅝-in. screws

# Building Cabinet Cases

As with the shelving cases described in chapter 8, I build cabinet units from ¾-in. hardwood plywood. The top, bottom, and sides of the basic box are cut from full sheets of plywood and are butt-joined together with #20 biscuits. As described in chapter 8 (see p. 130), I rabbet a scribe fin along the back edge of the sides that will be visible when installed, and this fin accepts the cabinet back. Sides that butt walls are installed with a scribe strip, or the face banding is scribed to fit tightly to the wall, as described in chapter 8.

## Base units

Base cabinets are built as simple boxes that sit on a kick base. Not all base units are the same depth; indeed, the great advantage of building your own cabinets is that you can construct them to fit the space available. Nevertheless, a common depth is 24 in., a dimension that makes the most efficient use of panel stock.

I begin by ripping the panel stock to width. If I am using plywood, I rip the width to 23⅞ in. (half the width of a 4-ft.-wide sheet, less the kerf). If I am using MDF, I can usually rip to a full 24 in., since the sheets often come 49 in. wide (see pp. 112–121 for panel stock options).

I prefer to build base units no wider than 4 ft., a size that corresponds to an even multiple of 8-ft.-long panel stock and that is still relatively manageable. Much bigger than that and the units

The sides of a fixed-shelf unit are prepped with a biscuit joiner. Note that a rabbet—a step that accepts the ¼-in. plywood back and leaves a narrow scribed fin—cuts along the back edge of each side piece.

## Top Treatments

Cabinet tops can be handled in two ways. If the cabinet extends all the way to the ceiling, you can build a fascia, as shown in the drawing on the facing page. If the cabinet doesn't extend to the ceiling, you can, of course, simply trim out the top of the case with a simple cove or bed molding (shown below). A design that calls for wrapping the top with wide crown molding requires sufficient backing. The easiest way to do this is to build a separate top that slips over the top of the case, as shown in the drawing on p. 154. This design also provides a useful display shelf on top of the cabinet.

A rabbet can be cut with a bearing-guided router bit.

## + SAFETY FIRST

Don't pick up a full sheet of plywood, especially a ¾-in. sheet, that is lying flat on sawhorses. That's a sure way to put out your back. Plywood is very heavy and a horizontal position is awkward. Instead, ease one edge down to the floor, and then pick up the sheet in the vertical position.

## PRO TIP

*A level base is critical when installing a cabinet case and even more so when installing multiple units.*

## IN DETAIL

CDs are 4⅞ in. wide and cassettes are 4¼ in. wide. I usually put both in drawers, dividing a standard 20¼-in. drawer into 5-in. rows. If you prefer to put them on a shelf, make the shelves at least 5⅛ in. tall; any lower and the cases become too difficult to tip out. Video and DVD cases are 7½ in. tall; they fit well in a standard drawer divided into thirds or on a shelf at least 8 in. tall.

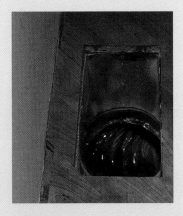

## IN DETAIL

An advantage of using a plywood fin base is that it allows in-floor heating vents to sit below the cabinet. The fins create a channel to redirect heat from the floor vent to a vent installed in the finished kick panel.

A long run of cabinets is built from several units that are joined together.

Stretchers cut from 4-in.-wide plywood stabilize the front and sides of the cabinet case while providing a screw base for the countertop.

become unwieldy. In general, when I lay out unit dimensions, I divide separate units into multiples of 2 ft. to conform to panel stock dimensions. However, when filling in an uneven length—say, 9 ft.—it often makes sense to build one 5-ft. and one 4-ft. unit. I try to keep unit lengths close to equal, and then build only one odd-size unit.

Cabinets wider than one set of double doors need a vertical divider to support doors in the middle of the run. These verticals are cut to the same dimension as a closed side panel and biscuit-joined to the bottom, as shown in the drawing on p. 150. Doors supported on fixed vertical dividers are hung on half-overlay hinges, while doors supported by end panels have full-overlay hinges.

Base units that will be covered by a countertop don't need a full-width top panel. Instead, I cut 3-in.- to 4-in.-wide "stretchers" to span the top of the cabinet, support the verticals, and provide a screw base for attaching the countertop.

## Cup Hinge Types
*With different size baseplates, cup hinges can accommodate a variety of cabinet door styles.*

**Full overlay**

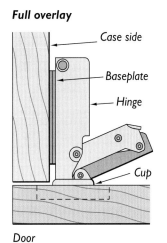

Case side — Baseplate — Hinge — Cup

Door

**Half overlay**

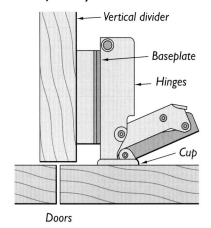

Vertical divider — Baseplate — Hinges — Cup

Doors

**Inset**

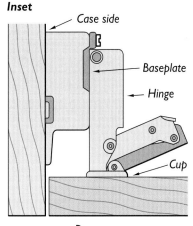

Case side — Baseplate — Hinge — Cup

Door

## Facing

Once the cases have been built, I band the front edges. For plywood cabinets, I sometimes use iron-on veneer tape (see p. 163) to hide the edge grain. However, the tape tends to crack and peel at the edges if it is banged frequently. A ¾-in.-thick solid hardwood face frame holds up better over time.

Face framing is similar to that described in chapter 8; however, it is attached to the cabinets before the units are installed. In this case, the unit doesn't need a fascia across the head. Instead, I glue and nail the vertical pieces so that they run continuously along the sides. These pieces are ⅞-in.-wide; they cover plywood edges and leave a small (about ⅛-in.) overhang that can be scribed and planed so the units fit tightly together and to end walls. I make sure that the facing is flush to the inside of the cabinet and overhangs to the outside only, oth-

**Shims below the plywood-fin kick base level the case. After the case is secured to the wall, the shims will be cut back and a finish kick panel will be installed over the front of the fins.**

erwise it may interfere with door hinges and drawer slides. Between vertical facings, I in-fill with 1½-in.-wide horizontal pieces along the bottom (flush with the top edge of the case's bottom panel) and 1¼-in.-wide horizontal pieces flush with the front stretcher (see the photo on p.152).

## Kick base

I build the kick base from 2×4s (as described in chapter 8) or as a series of plywood fins installed perpendicular to the face of the cabinet. If the floor is wildly out of level, I use a 2×4 "ladder," which I can level before installing the cabinet cases (see the drawing on p. 138). A level base is critical when installing any cabinet case and even more so when installing multiple units. If the units all sit on the same level base, it is easy to keep them square and aligned to each other. If the floor is level, however, the plywood fin approach is faster and more elegant. Instead of using pocket screws to mount the fins, you can build L-shaped brackets from small pieces of plywood.

Install fins or brackets every 16 in. to 24 in. o.c. to provide suitable backing for the kick panel and to provide plenty of cabinet support. A 36-in.-wide base cabinet, for example, typically has sup-

**Plywood fins can be quickly installed on a cabinet (here, on the base of a case-style window bench). Use a framing square to align the fins, which have been predrilled for pocket screws.**

**PRO TIP**

*When a shelf unit will fit below a slanted ceiling, it's wise to leave a bit more material to scribe-fit.*

## IN DETAIL

Knockdown (or KD) hardware was originally developed in Europe, where homeowners typically take their kitchen cabinets and other modular cabinetry with them when they move. One of the most useful pieces of KD hardware is the Titus® connector (see Resources on p. 167) for securing shelves. The connectors have studs that thread into shelf-peg holes, and the studs interlock with connectors inserted into 35mm holes (drilled with the same Forstner bit used for cup hinges) on the bottom face of the shelves. Turning the connector draws the stud in tight, strengthens the shelf-case connection, and eliminates the gap between the shelf and the case.

Screw adjacent cabinets together through predrilled holes in the edges of the face frame. Here, sliding clamps hold the cabinets tightly together until the screws are driven home.

After all the base units are attached to each other, secure the run snugly to the end walls.

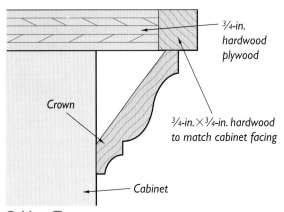

**Cabinet Top**
*To support wide crown on top of a cabinet, build a plywood cover that extends beyond the face.*

- ¾-in. hardwood plywood
- Crown
- ¾-in. × ¾-in. hardwood to match cabinet facing
- Cabinet

To quickly find the stud spacing, Clark Sargent pulls out his tape measure and lays it across the back of the cases.

ports beneath each end and in the center. Wider cabinets should have fins directly beneath vertical supports, as well.

## Securing bases

Before securing bases to the wall and floor, I usually screw adjacent base units together. Begin by planing the face frame for a tight fit between cabinets and against the end walls. Then predrill through the edge of the face frame and clamp it before sinking the screws.

Next, secure base units to the wall by driving 3-in. wood screws into each wall stud. If the wall is out of plumb or out of square relative to the cabinets, shim between the wall and the case or you may rack the cases out of square as you drive the screws home. Add a few judicious screws to hold the base. If the cabinet is on a ladder base, screw down through the bottom panel into the leveling base (which should already be level and secured to the floor). If the base units sit on a fin base, predrill the front edge and drive toe screws through the fins into the floor. Most bases are deep and not at risk of tipping over, so you don't need a lot of screws to secure a run of cabinets.

Shelf cabinets often have fixed vertical shelf dividers corresponding to those on the base units beneath them.

## Shelf units

I make shelf cases in units and at the same width as their bases so that they align vertically. As I've mentioned in chapter 8, I prefer to keep shelf spans short (under 2 ft. wide) to reduce the likelihood of the shelves bowing under the weight of heavy books or other items. This means that the shelf cabinets often need fixed vertical shelf dividers corresponding to those on the base units below. For verticals spaced no more than 24 in. apart, a ¼-in.-thick luaun or hardwood-faced plywood back provides plenty of strength.

For cabinets with adjustable shelves, I typically use the same shelf pins as those for bookcases (see p. 132). If the shelf unit is very tall, I usually try to install some fixed horizontal shelves for added strength. However, there are times when fixed shelves are inappropriate to the design of a large cabinet, so I use knockdown connectors instead of standard shelf pins because they are stronger and help reinforce the cabinet. (See In Detail on the facing page).

Before scribing, make sure that the unit is plumb.

When the unit is sitting plumb, scribe the face frame, holding a pencil flat to the ceiling.

Remove the cabinet and plane along the scribed line, canting the plane toward the back to create a slight back bevel on the face frame.

## WHAT CAN GO WRONG

Don't cut drawer pieces until you have built the cabinets. Plywood thicknesses vary, and cabinet pieces can shift slightly when the case is assembled. However, if you have already built the case, you can measure the exact opening required.

## IN DETAIL

The most common European-style slides are three-quarter extension models, meaning that three-quarters of a standard-depth drawer rolls out of the case. Full-extension slides, which allow the full depth of the drawer to roll out, are also available. Although more expensive, they may well be worth it because they offer access to additional storage.

## TRADE SECRET

Although today's epoxy-coated roller slides represent a huge improvement over the rattling metal slides of the past, most of the time I use the premium Accuride® 3017. This full-extension, all-ball-bearing slide is wonderfully simple to install.

---

The facing for open shelves is also a full frame made from ¾-in.-thick solid wood. This, too, provides a much needed scribe surface for seamlessly installing the units. As with the face frame on a base unit, I typically let ⅛ in. to ¼ in. of the face frame overhang the sides that butt unit to unit. When a shelf unit is installed below a slanted ceiling, it's wise to leave a bit more material to scribe the fit—at least ½ in. extra. Before scribing, make sure the unit is plumb, then scribe the unit to the wall and plane to fit (see the photos on p. 155).

# Drawer Construction

Most houses do not have enough drawers. That's because drawers take considerable time to build, so they are often used sparingly in a cabinet design. But where else can you conveniently store the myriad small items in your home? Even though they're time-consuming to build, it's worth including as many drawers as possible in a built-in.

Drawers also take a lot of abuse. Frequent opening and closing puts considerable strain on

### Drawer Dimensions

*To function properly, drawers must be built to precise dimensions. Here the dimensions for a drawer in a 24-in.-wide cabinet uses drawer slides that require 1 7/32 in. of clearance on each side of the drawer box.*

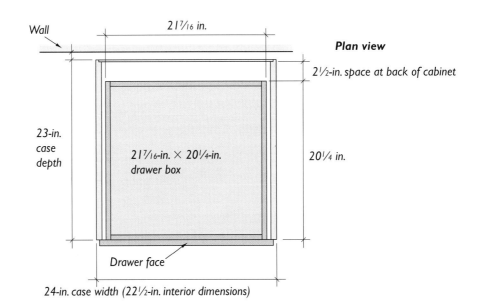

**Plan view**

Wall

21 7/16 in.

2½-in. space at back of cabinet

23-in. case depth

21 7/16-in. × 20¼-in. drawer box

20¼ in.

Drawer face

24-in. case width (22½-in. interior dimensions)

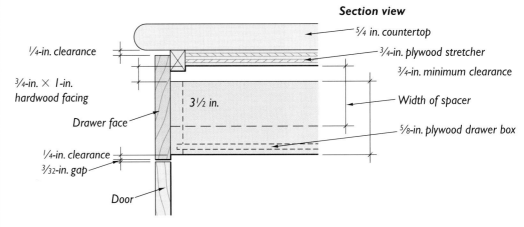

**Section view**

¼-in. clearance

¾-in. × 1-in. hardwood facing

Drawer face

¼-in. clearance
3/32-in. gap

Door

3½ in.

5/4 in. countertop

¾-in. plywood stretcher

¾-in. minimum clearance

Width of spacer

5/8-in. plywood drawer box

**Edge-Glued Countertop**
*When gluing boards edge to edge to make a wide countertop, pay close attention to the orientation of each board to prevent the final piece from warping.*

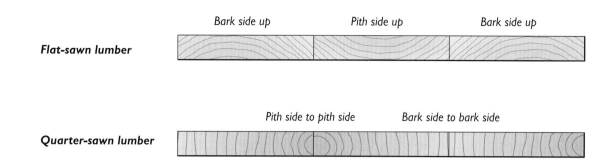

Flat-sawn lumber — Bark side up · Pith side up · Bark side up

Quarter-sawn lumber — Pith side to pith side · Bark side to bark side

the box joints and the slide. However, if you are patient cutting the pieces accurately, joining the box, and mounting the slides, a drawer will stay square and operate smoothly for a long time.

I use a simple but sturdy drawer design that can be reproduced fairly quickly. It's not the most elegant joinery, but once you've mastered the technique and understand the conditions necessary to make a drawer function well, you can enhance the design by cutting more elegant joints if you have the time, budget, and inclination.

## Dimensioning drawers

Drawers must be built to precise measurements. If the drawer box is built even $\frac{1}{16}$ in. too wide, the drawer may bind. Too narrow, and the slides won't fit together or the drawer will operate with so much slop that it's bound to fail over time.

At issue here is the close tolerances that drawer slides require. Different slides require different amounts of clearance on each side of the drawer, so make sure you buy the slides before you start the project. Many slides require $\frac{1}{2}$ in. of clearance on each side. I typically use European-style slides that require $\frac{17}{32}$ in. of clearance on each side, so each drawer must be $1\frac{1}{16}$ in. narrower than the cabinet opening.

The inside of a 24-in.-wide cabinet made from $\frac{3}{4}$-in.-thick panel stock should be 22½ in. wide. Therefore, to use the European-style drawer slides, the outside dimension of the drawer box must be 21⅞ in. wide (22½ in.

minus 1⅟₁₆ in. equals 21⅞ in.). (See the drawings on the facing page.)

A standard drawer depth for a 24-in.-deep cabinet is 20¼ in. This dimension works well with most three-quarter-extension slides, leaving an extra 2½ in. of empty space at the back of the cabinet for any wires and cables that may drop from a computer or stereo above. Even though you may not need this extra space, make sure you leave plenty of clearance at the back. A three-quarter-extension slide allows only three-quarters of a standard-depth drawer to roll out. Extra depth added to the drawer will not increase the opening but will be hidden drawer space left inside the cabinet, so there's no real convenience to adding drawer depth unless you plan to use full-extension slides.

Next, determine the height of the drawer box. This depends on the size of the drawer front you have designed for the face of the cabinet. The drawer face typically rises $\frac{3}{4}$ in. above the top edge of the drawer box and hangs down about $\frac{1}{4}$ in. below the bottom of the drawer box. For example, a standard 5-in.-high drawer face (a standard-size drawer face at the top of a base unit) requires a drawer box that is 4 in. high.

## Cutting drawer stock

I cut the bottom of drawers from ¼-in. hardwood plywood. I prefer smooth-faced birch with one A-grade face, oriented in the drawer with the grain running parallel to the sides. This slides into

## PRO TIP

*Many manufacturers do not include instructions with drawer hardware. Study the hardware first to figure out the position of screws before you drill.*

## IN DETAIL

Most roller-slide manufacturers offer self-closing models. With self-closing slides, the drawer rises as it comes out of the cabinet. As you close the drawer, it reaches a point where it runs slightly downhill and closes. Since the drawer rises as it comes out of the case, you need at least ¼ in. of clearance (possibly more, depending on the slide) between the drawer face and the countertop.

## TRADE SECRET

Cut a block of scrap plywood to space a drawer slide the correct distance from the top of the cabinet, as shown in the bottom left photo on the facing page. When installing a slide on a base cabinet, turn the case on its side, so that gravity works in your favor. Push the slide all the way to the front edge of the case, and then hold the slide against the block as you screw it in place.

a ¼-in. by ¼-in. groove cut into the front, back, and sides of the drawer. Cut the plywood bottom so that it is a ½ in. larger in each direction than the inside of the drawer box; this allows it to fit into the groove on all sides.

For the sides of my drawer boxes, I prefer to cut stock from ⅝-in.-thick Baltic birch plywood. I first rip the stock to the height of the drawer box (in this example, 3½ in.), and then rip the groove for the bottom ⅜ in. up from the bottom edge. A ¼-in. dado blade in a table saw works best for cutting this groove. A down-cut spiral or a compression router bit works just as well but requires more time to set up. As a third alternative, I just make two passes through the saw with

**The drawer bottom fits into ¼-in. grooves cut on all sides of the drawer box.**

a standard ⅛-in.-thick carbide blade. Run a trial groove before ripping all of the drawer stock, and test-fit the bottom in case the plywood thickness varies.

Next, cut the sides, front, and back of the drawer box to length. Cut the sides full length (20¼ in.) and the front and back 1¼ in. shorter than the total width (20 3⁄16 in.) so that they fit inside the two sides (see the drawing on p. 156).

When all of the stock has been cut, you are ready to build the drawers. I screw the sides to the front and back of the drawers with 1-in. wood screws. Predrill holes through both ends of each side. Attach the sides to the front piece first, creating a U-shape, and then slide the bottom into its groove and attach the back.

## Installing drawer slides

Most slides have a component that mounts in the cabinet and a component that mounts on the drawer. Usually, rollers are captured in a U-channel on one component; the other component has a flat surface on which the rollers move.

To mount the slide on the drawer box, turn the drawer upside down on your work surface. Position the drawer component on the bottom edge of the drawer, flush with the front. Attach the slide with #6 pan-head screws, which won't interfere with the drawer's track. Elongated holes in the slides allow you a small amount of vertical adjustment. At this point, drive screws only in the center of the elongated holes so that you can adjust the drawer position later.

To mount the corresponding component in the cabinet, it helps to turn the entire case on its side. Measure the position of the slide on the side of the drawer box and add ¾-in. for clearance at the top. To simplify this installation, cut a spacer from scrap plywood to use for positioning the slide on the side of the cabinet. On the cabinet piece, elongated holes allow for adjustments back

**Predrill four holes in the drawer box to hold the drawer front.**

The last step is to check the travel of the slide and the reveal in the drawer front, making any final adjustments that may be needed. If the drawer rolls smoothly, no adjustments are necessary. But if the drawer bumps in its travel or jams, the slide positions need to be adjusted. Loosen the screws in the drawer channel and move it up or down, or loosen the screws in the cabinet channel and move it back and forth. When the slide travels freely without bumping or jamming, drive screws in two round holes in each channel. You don't need to lock in all of the screw holes; keep a few clear in case adjustments need to be made after the first heating season.

# Countertops

Much of the time, a countertop for a built-in cabinet functions more like a mantle or a decorative shelf, not like the laminate countertops that you find in a kitchen. It often serves as the bottom of upper shelves that sit on a bank of door-covered cabinets.

Because this type of countertop doesn't get the wear and tear that typically occurs in a kitchen, it is a good place to use a handsome piece of solid wood with a natural finish. In this case, it can

and forth. Again, only mount the slide with screws in the center of the elongated adjustment holes.

After the cabinet cases have been installed, install the drawer fronts. I begin by predrilling holes in the rough drawer boxes for four screws to go through the front panel, and then slip the rough drawer boxes into the slides. I then hold the drawer front in place, often using a scrap piece of 1× to gauge the ¾-in. clearance at the top of the drawer box. When the drawer front is positioned correctly, I drive the screws home.

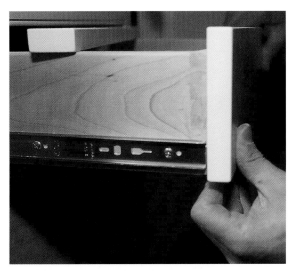

**The piece of 1× scrap lying across this drawer box was used to align the height of the drawer front.**

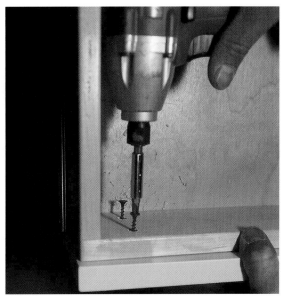

**Four 1¼-in. screws hold the drawer front in place. Before driving the screws, make sure that the drawer front is centered, with the same overhang on both sides.**

## PRO TIP

*If a bit has a lead point, test the depth on scrap wood before drilling any doors—you don't want to poke the point through the door face.*

This long countertop has been glued up from several pieces of 5/4 by 4-in. mahogany and given a bullnose profile along the front edge with a router-mounted roundover bit.

## IN DETAIL

As the name implies, a cup hinge is installed inside a "cup," or a 35mm round mortise bored into the face of the cabinet door. The carbide bits used to bore the hinges are expensive, but they are precise instruments. The Forstner-style bit severs the fibers at the perimeter of the mortise, while also cutting with a sharp knife-edge along the bottom. The result is a clean, flat-bottomed hole with no tearout at the surface.

## IN DETAIL

Allow a 1-in. to 1¼-in. overhang on the open ends of countertops that don't butt into walls. A flush end is difficult to align perfectly with the cabinet end, especially if the countertop must be scribed to fit.

serve as a bold accent that visually ties the built-ins to wainscoting and handrails, as the countertop in the balcony built-in does (see p. 148).

## Attaching the countertop

I typically install the countertop after the base cabinets have been set and secured to the wall. If the countertop fits into a corner, don't assume the wall is square—it rarely is. Often the end must be scribed and cut at a slight angle to fit against the walls. I begin by checking to see how square the wall is in relation to the cabinet, using a framing square and measuring along the front and back edges of the cabinet. The goal is to find the longest dimension, and then cut the countertop square to that length.

Set the countertop in place, scribing it to fit, if necessary, along the back and sides. When scribing

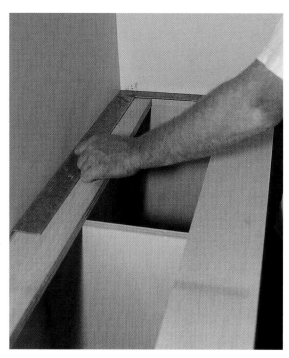

Use a framing square to evaluate the squareness of the wall corner in relation to the cabinet back before cutting the end of the countertop.

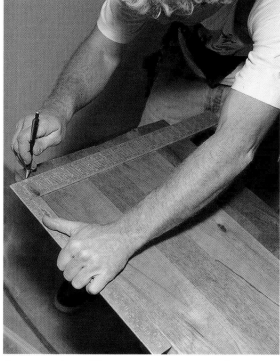

Carpenter Larz Allen squares up the end of a solid-wood countertop with a framing square.

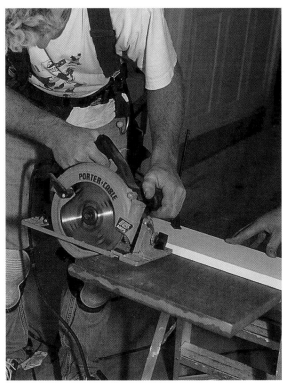

**For a straight cut, clamp a straightedge to the end of the countertop.**

**When scribing a countertop or bench seat to an end wall, hold the pencil flat to the wall and maintain a consistent angle while drawing across the board.**

a countertop or a bench seat to an end wall, I often use a pencil, holding it at a consistent angle along the wall and drawing across the board.

If an upper unit will be installed on top of the counter, only the ends that butt into a wall need to be scribe-fitted; any gap behind the countertop is usually covered by the back of the upper cabinet. I secure the countertop with 1¼-in. screws (for a ¾-in. top) or 1⅝-in. screws (for a 5/4 top), driven every 12 in. to 16 in. along the front and back ledgers on the base cabinets.

## ✚ SAFETY FIRST

When using a power sander, be sure to wear a dust mask. Fine wood particles lodged in your lungs can produce a nasty cough and cause serious damage over time. One option is to purchase a sander that collects dust and routes it into a bag or a shop vacuum. Even so, you should still wear a dust mask.

# Doors

With European-style cup hinges, the doors overlay the edges of the cabinet case, rather than being inset in the openings of the face frame. This style is easier and faster to build, and because the hinges are adjustable, the doors operate smoothly with minimal effort (see Essential door hardware, pp. 164–166).

A more traditional look requires building frame-and-panel doors. As with wainscoting panels, real raised-panel doors are time consuming to build. If I go this route, I sometimes make a simplified version that is relatively fast and easy to construct (see the sidebar on p. 162). Most of the time, however, I order true raised-panel doors that are manufactured to my specifications (see Resources on p. 167).

In its simplest incarnation, an overlay door is nothing more than a slab of hardwood plywood banded around the edges. Drawer fronts can be fashioned from the same materials, so the front of

# SIMPLIFIED FRAME-AND-PANEL DOOR CONSTRUCTION

Frame-and-panel doors were invented to solve the problem of wood movement. A solid-wood panel pinned in a frame will soon crack. In frame-and-panel construction, the panel "floats" in the frame, expanding and contracting as necessary. I make fill panels out of MDF, which is very stable and doesn't expand and contract with seasonal changes in humidity. Therefore, my frame-and-panel doors are decorative rather than practical.

My modified design uses a 1×4 frame made of maple or poplar and an MDF fill panel. I begin by ripping 1×6 poplar in half to get 2¾-in.-wide rail and stile stock. Then I cut a ½-in. by ½-in. rabbet along the milled edges of each board. I butt-join the pieces with the rabbet facing the inside of the frame and the ripped edge on the outer perimeter. I usually build frames approximately ½ in. larger in both directions and then trim them to the exact size on a table saw, using a jointer or a plane (hand or power) to smooth the cut edges. I'm careful not to cut the exposed face of the rails less than 1¾ in. wide; otherwise, cup hinges won't fit. When joining the stiles and rails, I cut a rabbet in the ends of the stiles that matches the rabbet on the rails. Although I use a table saw to cut the rabbet on the rails, I typically crosscut the rabbet on the stiles with a power miter saw equipped with a depth stop.

The last step is to cut a fill panel from ⅝-in.-thick MDF. I cut the panel ¼ in. undersize and rout a small cove along the perimeter. This panel "floats" in the frame and is secured with a light bead of adhesive caulk and ½-in. brads nailed from the back through the rabbet.

**To cut a door to its precise dimensions, trim it to the exact size on a table saw, and then smooth the cut edges by running them through a jointer, as shown here.**

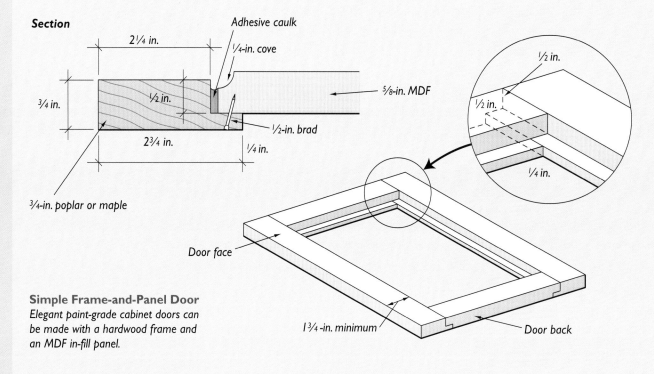

*Section*

Adhesive caulk

2¼ in.

¼-in. cove

⅝-in. MDF

¾ in.

½ in.

½-in. brad

2¾ in.

¼ in.

½ in.

½ in.

¼ in.

¾-in. poplar or maple

Door face

**Simple Frame-and-Panel Door**
*Elegant paint-grade cabinet doors can be made with a hardwood frame and an MDF in-fill panel.*

1¾-in. minimum

Door back

the cabinet becomes a smooth, seamless plane with a ³⁄₁₆-in. gap between the doors and the drawer fronts.

## Banding

Plywood doors must be banded along the edges to hide the edge grain. The easiest way to do this is to use iron-on banding, or veneer tape. While this material is relatively inexpensive and easy to install, it is fairly delicate and often starts to separate after a few years of use, particularly if the edges are knocked or pulled. More often, I apply iron-on banding to the edges of the cabinet case behind the doors and use ¼-in. hardwood edging for the doors. Solid-wood banding takes time to apply well, but it is very durable over the long haul. I also like the look of a painted wood door with a natural wood accent around the edges.

**Veneer tape.** This hardwood veneer with heat-sensitive adhesive comes in rolls slightly wider than the edges of ¾-in. plywood. It can be applied with a banding iron available from some woodworking supply houses or with a normal household iron (I bought one at a yard sale exclusively for this use) on dry (no steam).

Begin by cutting lengths of veneer tape about 1 in. longer than the width of all the doors. Hold it on the edge and press down firmly with the iron until the glue softens (you can see it bubble around the edges). If you move too slowly, you can scorch the veneer; too fast, and the tape peels off. With practice, you'll find the right pace.

Once the tape is adhered tightly, flip the door over and use a utility knife to trim the extra length flush with the ends of the door. I apply the tape to each side first, trim it, and then apply it to the top and bottom edges.

**Solid-wood edging.** I prefer to cut solid edging from ¼-in.-thick strips of hardwood. I first joint the edge to smooth off any saw marks, and then run the board through the table saw. This

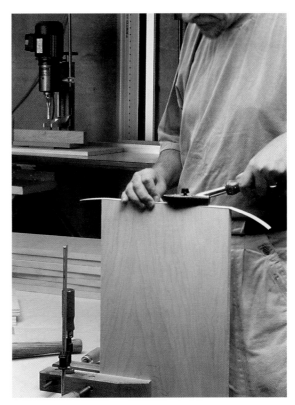

Apply veneer tape with a hot iron. After the adhesive cools, cut the ends of the tape flush to the corners with a utility knife and plane the edges flush to the faces.

gives me strips with one smooth edge that faces out and one sawn edge that faces the door edge. I prefer to apply these strips along the top and bottom edges of the door first, trim them to the width of the door, and then run the side banding continuously over the ends of the top and bottom edges. When applying the strips, use yellow carpenter's glue and nail every 6 in. with ½-in. to ¾-in. brads.

## Cutting doors

The first thing I do when making solid plywood doors is calculate the size of the doors and drawer fronts. For the door width, I measure the width of the cabinet case (outside to outside) and subtract ³⁄₈ in. (³⁄₁₆ in. is the width of the gap between doors) from each side of each door. Then I subtract for the thickness of the banding—³⁄₃₂ in. for veneer tape and ¼ in. for hardwood. Similarly, for

## IN DETAIL

Traditional face-frame cabinets with full inset doors mounted on leaf hinges are elegant but time-consuming to build. You must use frame-and-panel doors (leaf hinges don't work on the edge grain of plywood). Also, if a door warps, the case settles, or a child swings on the door and springs the hinges, you'll likely have to reset the door. European cup hinges, on the other hand, are much stronger and can be readjusted with a screwdriver in just a few minutes.

## TRADE SECRET

Installing 35mm European-style hinges requires perfect alignment between the cup and the screw holes on the door and between the baseplate and the screw holes on the cabinet. Laying out the hardware can be a painstaking process if you don't use a hinge template. I have found the Jig-It templates (from Rockler<sup>SM</sup>) to be a good system for accurately locating and boring cup holes and aligning the baseplate on the cabinet. Rockler offers jigs and the proper-sized bits for a number of different hinges as well.

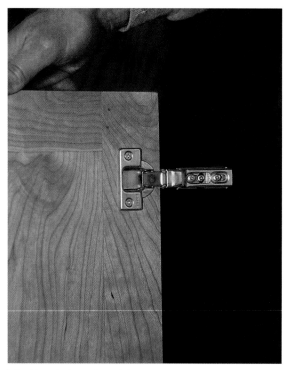

The "cup" side of the hinge mounts in a 35mm round mortise bored into the inside face of the door.

The baseplate mounts to the side of the cabinet at a precise distance from the top and front edges of the case.

the door height, allow for a ³⁄₁₆-in. gap between the top of the door and the bottom of the countertop or drawer face, and don't forget to subtract for the banding thickness when chopping the door to height.

Often the trickiest part of cutting doors comes in the planning stages. Determine the plywood division, so that you can get the most out of each sheet of plywood (for more on cutting plywood efficiently, see the drawing on p. 15). When I know the plywood division, I rough-cut the door into manageable slabs with a straightedge and a circular saw, and then cut to the final dimensions on a table saw.

## Essential door hardware

Cup hinges are the secret behind frameless built-ins. These innovative "Euro-style" hinges remain completely hidden when cabinet doors are closed, giving the cabinet front an uninterrupted, uncluttered face. Compared to leaf hinges, cup hinges make hanging cabinet doors much easier, as long as you have the right tools. The biggest advantage is that cup hinges can be adjusted after they have been installed, allowing you to evenly space the gaps between the doors. Cup hinges work equally well with all door styles, and some even work on face-frame cabinets.

Cup hinges have two parts: a cup, which mounts on the inside face of the door, and a baseplate, which screws to the side of the cabinet case. The weight of the door is distributed over the entire rim of the cup, which makes for a relatively strong hinge compared to a traditional leaf hinge, which puts all of the door's weight (and that of a child who might swing on it) on just a few screws. The collapsible arm attached to the cup slides onto a baseplate that can be adjusted about ⅛ in. in three directions: up, down, and sideways. Conventional leaf hinges have no adjustment capability, so you not only need to be dead on when you install them, but you also have no recourse for readjusting the hinge over time.

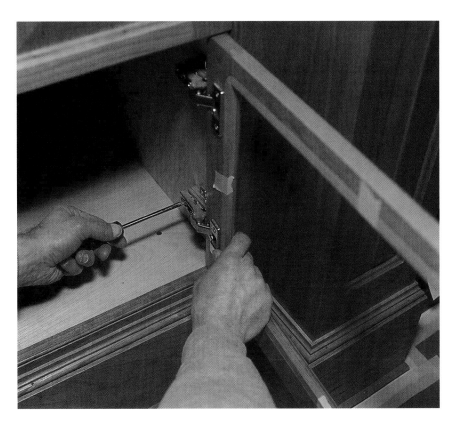

Adjust cup hinges with a Pozi-drive screwdriver to make the gap between the edge of the door and the rail or adjacent door equal and parallel.

**Mounting cup hinges.** Installing cup hinges is a two-step process: The cup, which holds the hinge arm, is mounted on the door and the baseplate is mounted on the case. Drill a 35mm-diameter hole for the hinge using a special Forstner bit. Most of these are carbide—high-speed steel doesn't hold up very long when boring into MDF and plywood. The hole should be drilled to the depth of the bit head—about ½ in. Not all bits have a lead point, so unless you have a drill press, you will need to make a jig to guide the bit.

Mounting the baseplate requires careful positioning on the cabinet side. You can make your own jig for this or buy one from the hinge manufacturer. To make the job easier, tip the cabinet on its side, so gravity works in your favor. Use #6 ⅝-in. oval-head sheet-metal screws. Unlike wood screws, sheet-metal screws have a consistent shaft diameter, so they maintain a better grip in MDF and particleboard.

Some hinges come with 7.5mm "Euro" screws, a larger-diameter screw designed specifically for

## Cup Hinges

Cup hinges are designed to mount on the inside face of a cabinet door. The most common configuration for full-overlay doors allows the doors to open to an angle of 110 degrees to 130 degrees. Cup hinges can usually be converted to work for half-overlay doors by using a spacer, or for inset doors by using a new baseplate. A wide variety of specialty types are also available for doors that swing up to 175 degrees, as well as for extra-thick doors, blind-corner doors, and glass doors.

When choosing the opening angle of the hinge, consider where the cabinet is placed. For wall cabinets, a 120-degree hinge works best, allowing an unobstructed opening of the entire cabinet. For base cabinets, I use hinges that open to 165 degrees. This permits the door to open almost flat against adjacent cabinets and prevents the door from being ripped out if it's left open and someone runs into it. For base cabinets against a wall, I use hinges that open to 110 degrees to prevent the door from hitting the wall.

Major cup hinge manufacturers have added a snap-on feature that allows you to mount or dismount the door without tools. You simply pull a small lever to release the door. This means that when it's time for a plumber to replace a sink trap or service the disposal, you can quickly remove the sink cabinet doors and set them aside.

## WHAT CAN GO WRONG

Check television and stereo dimensions carefully before building a cabinet. Include a couple inches of breathing room on each side of the equipment to help you slide the unit in and out when setting it up or connecting a VCR.

## TRADE SECRET

I've learned that a fully tooled industrial shop can produce better doors and drawer fronts (see Resources on the facing page) for less money and in less time than I can on site. I usually order doors and drawer fronts from the Cabinet Factory® in LaCrosse, Wisconsin, though there are several other manufacturers, too. Manufacturers offer these items in a number of wood species, styles, and configurations. They also prebore for cup hinges and deliver in two to three weeks.

## Tips for Using Cup Hinges

- Always test the hinge placement by mounting both parts of the hinge on two pieces of scrap plywood before mounting the hardware on a cabinet.
- Make the height of doors greater than the width.
- Keep hinges as close as possible to the top and bottom edges of the doors—a maximum of 3 in. is recommended
- Don't exceed the maximum cup drilling distance from the edge of the door. This will cause the doors to bind.

- Attach the baseplate using all available screw holes—a great deal of force is applied to the plate when the door is opened and closed.
- Make sure that the hinge arm is square to the edge of the door. Improper mounting makes it harder to open the door and could eventually cause the hinge to fail.
- Don't overuse hinge adjustments. If you need to make excessive adjustments, check your mounting jigs and measurements.

particleboard. Or they may come with nylon inserts that fit into even larger holes and accept a standard #6 screw. Nylon inserts are probably the best option for particleboard, because they allow the screw to be removed and replaced many times.

Every brand of hinge has its own method of adjustment. Some hinges are adjustable in two dimensions, others in all three dimensions. The screw that is the farthest back adjusts the height, so loosen it to move the hinge up or down. The middle screw enables you to adjust the door in or out to align the front face of the doors. The outermost screw adjusts the hinge sideways to equalize the gap between adjacent doors. Loosen it to move the door away from the baseplate; tighten it to move the door toward the baseplate.

# Resources

## Abrasives

Klingspor's Woodworking Shop
856 21st St. SE
P.O. Box 5069
Hickory, NC 28603
(800) 228-0000
www.woodworkingshop.com

## Bosch® DWM40L Miter Finder

S-B Power Tool Company
4300 W. Peterson Avenue
Chicago, IL 60646
(877) 267-2499
www.sbpt-hr.com

## Cabinet Doors

Cabinet Factory
P.O. Box 1748
La Crosse, WI 54602
(800) 237-1326

## Collins Coping Foot

Collins Quality Tool® Company
P.O. Box 417
Plain City, OH 43064
(888) 838-8988
www.collinstool.com

## Color Putty®

Color Putty® Company
121 W. 7th Street
Monroe, WI 53566
(608) 325-6033
www.colorputty.com

## Cup Hinges and Drawer Slides

Grass America
P.O. Box 1019
Kernersville, NC 27284
(800) 334-3512

Hettich American L.P.
1607 Anaconda Road
Harrisonville, MO 64701
(800) 438-8424

Julius Blum Inc.
Blum Industrial Park
Highway 16 - Lowesville
Stanley, NC 28164
(800) 438-6785

Knapp and Vogt Mfgr. Co.
2700 Oak Industrial Drive NE
Grand Rapids, MI 49505
(616) 459-3311

Mepla
909 West Market Center Drive
High Point, NC 27261
(910) 883-7121

## Drawer Slides Only

Accuride
12311 Shoemaker Avenue
Sante Fe Springs, CA 90670
(310) 903-0200

## Hand Tools

Garrett Wade Co., Inc.
161 Avenue of the Americas
New York, NY 10013
(800) 221-2942
www.garrettwade.com

Highland Hardware℠
1045 N. Highland Avenue, N.E.
Atlanta, GA 30306-3592
(800) 241-6748
www.highlandhardware.com

Lee Valley Tools
P.O. Box 1780
Ogdensburg, NY 13669-6780
(800) 871-8158
www.leevalley.com

Woodcraft®
560 Airport Industrial Park
P.O. Box 1686
Parkersburg, WV 26102-1686
(800) 225-1153
www.woodcraft.com

## Power Tool Accessories

Coastal Tool & Supply
510 New Park Avenue
West Hartford, CT 06116
(877) 551-8665
www.coastaltool.com

Hartville Tool®
13163 Market Avenue North
Hartville, OH 44632
(800) 345-2396
www.hartvilletool.com

Rockler℠
4365 Willow Drive
Medina MN 55340-9701
(800) 279-4441
www.rockler.com

Roto Zip® Tool Corp.
1861 Ludden Drive
Cross Plains, WI 53528
(877) ROTOZIP
www.rotozip.com

Woodhaven
501 West 1st Avenue
Durant, IA 52747-9729
(800) 344-6657
www.woodhaven.com

Woodworker's Supply, Inc.℠
1108 North Glenn Road
Casper, WY 82601-1698
(800) 645-9292

## Saw Blades, Router Bits, and Accessories

Amana Tool® Corporation
120 Carolyn Boulevard
Farmingdale, NY 11735
(800) 445-0077
www.amanatool.com

CMT® USA, Inc.
307-F Pomona Drive
Greensboro, NC 27407
(888) 268-2487
www.cmtusa.com

Eagle America℠
Chardon, OH 44024
(800) 872-2511
www.eagle-america.com

Jesada Tools®
310 Mears Boulevard
Oldsmar, FL 34677-3047
(800) 531-5559
www.jesada.com

# Index